POLLUTION

Other books in the Current Controversies Series:

The AIDS Crisis
Alcoholism
Drug Trafficking
Energy Alternatives
Europe
Free Speech
Gun Control
Illegal Immigration
Iraq
Nationalism and Ethnic Conflict
Police Brutality
Sexual Harassment
Violence Against Women
Women in the Military
Youth Violence

POLLUTION

David L. Bender, *Publisher*
Bruno Leone, *Executive Editor*

Bonnie Szumski, *Managing Editor*
Katie de Koster, *Senior Editor*

Charles P. Cozic, *Book Editor*

Cover photo: © Robert Reiff 1992/FPG

Library of Congress Cataloging-in-Publication Data

Pollution / book editor, Charles P. Cozic.
p. cm. — (Current controversies)
Includes bibliographical references and index.
ISBN 1-56510-076-X (lib. bdg. : alk. paper) — ISBN 1-56510-075-1 (pbk. : alk. paper)
1. Pollution. 2. Environmental protection. [1. Pollution. 2. Environmental protection.] I. Cozic, Charles P., 1957- . II. Series.
TD176.P65 1994
363.73—dc20 93-4552
CIP

AC

Printed in the U.S.A.

Contents

Foreword

By definition, controversies are "discussions of questions in which opposing opinions clash" (Webster's Twentieth Century Dictionary Unabridged). Few would deny that controversies are a pervasive part of the human condition and exist on virtually every level of human enterprise. Controversies transpire between individuals and among groups, within nations and between nations. Controversies supply the grist necessary for progress by providing challenges and challengers to the status quo. They also create atmospheres where strife and warfare can flourish. A world without controversies would be a peaceful world; but it also would be, by and large, static and prosaic.

The Series' Purpose

The purpose of the Current Controversies series is to explore many of the social, political, and economic controversies dominating the national and international scenes today. Titles selected for inclusion in the series are highly focused and specific. For example, from the larger category of criminal justice, Current Controversies deals with specific topics such as police brutality, gun control, white collar crime, and others. The debates in Current Controversies also are presented in a useful, timeless fashion. Articles and book excerpts included in each title are selected if they contribute valuable, long-range ideas to the overall debate. And wherever possible, current information is enhanced with historical documents and other relevant materials. Thus, while individual titles are current in focus, every effort is made to ensure that they will not become quickly outdated. Books in the Current Controversies series will remain important resources for librarians, teachers, and students for many years.

In addition to keeping the titles focused and specific, great care is taken in the editorial format of each book in the series. Book introductions and chapter prefaces are offered to provide background material for readers. Chapters are organized around several key questions that are answered with diverse opinions representing all points on the political spectrum. Materials in each chapter include opinions in which authors clearly disagree as well as alternative opinions in which authors may agree on a broader issue but disagree on the possible solutions. In this way, the content of each volume in Current Controversies mirrors the mosaic of opinions encountered in society. Readers will quickly realize that there are many viable answers to these complex issues. By questioning each author's conclusions, students and casual readers can begin to develop the critical thinking skills so important to evaluating opinionated material.

Current Controversies is also ideal for controlled research. Each anthology in the series is composed of primary sources taken from a wide gamut of informational categories including periodicals, newspapers, books, United States and foreign government documents, and the publications of private and public organizations. Readers will find factual support for reports, debates, and research papers covering all areas of important issues. In addition, an annotated table of contents, an index, a book and periodical bibliography, and a list of organizations to contact are included in each book to expedite further research.

Perhaps more than ever before in history, people are confronted with diverse and contradictory information. During the Persian Gulf War, for example, the public was not only treated to minute-to-minute coverage of the war, it was also inundated with critiques of the coverage and countless analyses of the factors motivating U.S. involvement. Being able to sort through the plethora of opinions accompanying today's major issues, and to draw one's own conclusions, can be a complicated and frustrating struggle. It is the editors' hope that Current Controversies will help readers with this struggle.

"Criminal penalties for toxic polluters have never been stronger."

Introduction

Amidst growing concern over polluted air, land, and water, America is cracking down on environmental criminals. At both the federal and state levels, criminal penalties for toxic polluters have never been stronger. Indeed, many infractions that were merely civil violations or misdemeanors only a few years ago are now prosecuted as felonies. For example, in the past, society believed that the environment was more resilient, and polluters expected nothing more than minor punishment, if any at all. But today, "midnight dumpers"—people who deliberately and secretly dump hazardous cyanide, mercury, and other toxic waste on land or into sewers—are serving three-year prison sentences. Despite this crackdown on toxic dumping, legal experts disagree as to whether prosecutors are pursuing other suspected polluters too aggressively or not aggressively enough. Experts on both sides believe that in many pollution cases, serious abuses of justice are taking place.

Consider the case of PureGro Company. According to *Business Week* writer Catherine Yang, for five years workers at the chemicals manufacturer "had disposed of leftover pesticides in an open tank in [Washington state]—without keeping records of what was dumped." Yang reported that in 1987, PureGro sprayed the chemicals on a leased cornfield to avoid classifying them as hazardous waste. In a 1992 report, Jonathan Turley, director of the Environmental Crimes Project at George Washington University, wrote, "Over twenty people suffered symptoms from exposure to the fumes from the dumped material. Jack Downs [a rancher living next to the cornfield] died approximately one year later. He spent most of his last year hospitalized." Turley said those afflicted had suffered severe health problems, including blistering lesions and respiratory and heart ailments. In 1990, PureGro was charged with illegally dumping hazardous chemicals, a felony under federal law. PureGro offered to plead guilty to one felony and pay a large fine, but the U.S. Department of Justice refused the plea bargain, and a grand jury indicted the company and four executives on five felony counts. However, Turley said that prior to trial "the Bush administration apparently felt that this was unfair and accepted a plea to a small misdemeanor and a $15,000 fine." He charges that corporate and political influence forced the settlement.

In 1993, Turley called the PureGro case one of many "identified by Congress that follow a pattern of reluctant prosecution by the Justice Department." Peter Montague, director of the Environmental Research Foundation, believes that corporate influence undermines government's prosecution of polluters. He accuses, "Government routinely uses its discretion to plea bargain with polluters, letting poisoners off easy. Given a choice, government refuses to enforce antipollution laws."

Other legal experts, however, argue that federal and state prosecution is hardly weak, but, rather, overly rigorous. While many welcome the criminal prosecution of midnight dumpers, these experts contend that federal law grants prosecutors such vast power and discretion in choosing which other violators to prosecute as felons that any violation—no matter how minor or accidental—can be construed as a crime and could result in a mandatory prison term. As Judson Starr, an attorney and former head of the Justice Department's Environmental Crimes Section, explains, "The government does not have to prove that the act was done with an evil purpose. A prosecutor could prove a felony merely by showing that the defendant knew that he or she was doing the physical act itself."

One case illustrating such overzealousness, according to *Forbes* writer Leslie Spencer, involved Diceon Electronics, a California manufacturer. In 1990, state prosecutors charged Roland Matthews, chief executive officer of Diceon, and two other company officers with illegally storing hazardous waste in leaky containers that were not labeled according to regulations. According to Spencer, "No one was hurt. No premeditation was charged." Nevertheless, the two officers were put on probation and the company was fined $600,000. Although criminal charges against Matthews were eventually dropped, an appeals court refused to have his arrest records sealed and destroyed. Starr criticizes prosecutors in such cases for seeking felony convictions, which could ruin a company financially or destroy an individual's reputation. He feels such penalties far outweigh their crimes and are contrary to what the goals of justice should be. "The criminal program," he states, "should be a very powerful deterrent but used very sparingly."

Advocates of a get-tough policy argue that harsh penalties, including imprisonment, are the only message polluters will heed. But critics contend that the threat of such penalties alone is a sufficient deterrent. Since mandatory sentencing guidelines now force many judges to impose imprisonment or other severe punishment, seeking felony convictions is a serious step that prosecutors must consider carefully. As government strives for both environmental protection and justice, the issue of how to prosecute environmental crimes promises to endure for many years to come.

Pollution: Current Controversies focuses on two important aspects of the environmental crime debate: Are corporations major polluters and is government adequately protecting the environment? Other viewpoints in this anthology discuss recycling and the seriousness of pollution. These viewpoints will increase readers' understanding of not only corporate and government roles regarding pollution, but also the physical effects of pollution and the economic concerns related to cleaning up the environment.

Chapter 1

How Serious a Problem Is Air Pollution?

Chapter Preface

In many large cities, automobiles and factories pollute the air so heavily that smog endangers the residents' health. For example, in Los Angeles and Mexico City, where smog is a constant problem, children, the elderly, and people with asthma or heart problems are often struck ill or incapacitated from breathing dirty air. While these health problems are all too real, the harm from another form of air pollution, greenhouse gases, is less clear and hotly debated.

Greenhouse gases include carbon dioxide, chlorofluorocarbons (CFCs), and methane, which rise miles into the atmosphere, far above the smog that hovers over cities. The effects of these greenhouse gases on phenomena such as ozone layer depletion are complex and scientists have only begun studying them relatively recently.

In 1974, for example, University of California at Irvine scientists Sherwood Rowland and Mario Molina theorized that CFCs deplete the ozone layer, a stratospheric shield approximately ten to thirty miles above earth that blocks out the most harmful effects of the sun's ultraviolet (UV) radiation. Many scientists argue that this depletion, a virtual hole in some places, is increasing and raising the risk of skin cancer and weakening humans' immune systems. Other experts, however, disagree and argue that nature continuously creates and destroys ozone, so any depletion is merely temporary. Scientist and former Washington governor Dixy Lee Ray agrees and adds, "There is no documented evidence of CFC molecules in the stratosphere. There is no overall loss of ozone."

Despite the unanswered questions, the United States and other nations have agreed to ban the production of CFCs. However, industrial activity such as burning fossil fuels continues to pump greenhouse gases into the atmosphere. While some believe that immediate action is necessary to avert future problems, many scientists believe that more time is needed to study the effects of greenhouse gases on the ozone layer. The viewpoints in the following chapter discuss the threat of this and other forms of air pollution.

Air Pollution Is a Global Crisis

by Al Gore

About the author: *Al Gore served as a U.S. senator for seven years before he became the forty-fifth vice president of the United States in 1993. Gore is the author of* Earth in the Balance: Ecology and the Human Spirit, *from which this viewpoint is excerpted.*

The magnitude of the changes we are imposing on the world's climate pattern is made obvious from the perspective of history, but in any given year our attention is likely to focus on the swirl of contemporary events—and specific problems with pollution, particularly of the air. No sooner had the political dust of Eastern Europe's revolution against communism settled in 1989 than the world gasped in horror at the unbelievable levels of pollution—especially air pollution—throughout the communist world. We learned, for example, that in some areas of Poland, children are regularly taken underground into deep mines to gain some respite from the buildup of gases and pollution of all sorts in the air. One can almost imagine their teachers emerging tentatively from the mine, carrying canaries to warn the children when it's no longer safe for them to stay above the ground.

One visitor to the Romanian "black town" of Copsa Mica noted that "the trees and grass are so stained by soot that they look as if they had been soaked in ink." A local doctor reported that even horses can stay only for two years in the town; "then they have to be taken away, or else they will die."

In the northern reaches of Czechoslovakia, the air is so badly polluted that the government actually pays a bonus to anyone who will live there for more than ten years; those who take it call it burial money. To the east, in the Ukraine, that one republic puts eight times as many particulates into the air each year as does the entire United States of America.

Throughout the developing world, similar nightmares are found on every continent. In Ulan Bator in Outer Mongolia, the local beverage, curdled mare's milk, has to be protected from the black flecks in the air that settle on every surface. Mexico City suffers the worst air pollution, day in and day out, of any city on earth. There are occasional tragedies, such as the accidental release of poisonous gas into the air above Bhopal, India, which capture the whole world's attention. But the constant deadly levels of air pollution in cities throughout the developing world do not, even though on a "normal" day they are responsible for the deaths of more people than died at Bhopal.

> ***"[The Ukraine] republic puts eight times as many particulates into the air each year as does the entire United States of America."***

The developed world, including the United States and Japan, has its own problems with air pollution, of course, in cities like Los Angeles and Tokyo. But there have also been some dramatic successes. Pittsburgh, once legendary for its thick, soupy air, is now one of the most livable cities in the world. Most residents of Nashville don't even know that their city was once called Smokey Joe. London still has serious problems, but nothing that compares with the "killer smogs" of the 1950s. And since the Atmospheric Nuclear Test Ban Treaty stopped most aboveground nuclear explosions in the 1960s, the levels of deadly strontium 90 in the air have fallen dramatically.

Some of the successes in dealing with air quality have created new problems. For example, the use of tall smokestacks to reduce local air pollution has helped to worsen regional problems like acid rain. The higher the air pollution, the farther it travels from its source. Some of what used to be Pittsburgh's smoke is now Labrador's acidic snow. Some of what Londoners used to curse as smog now burns the leaves of Scandinavian trees.

And while many of the measures that control local and regional air pollution also help reduce the global threat, many others actually increase that threat. For example, energy-consuming "scrubbers," used to control acid precipitation, now cause the release of even more carbon dioxide (CO_2) into the atmosphere. A power plant fitted with scrubbers will produce approximately 6 percent more global air pollution in the form of CO_2 for each BTU [British Thermal Unit] of energy generated. Moreover, the sulfur emissions from coal plants partly offset, and temporarily conceal, the regional effects of the global warming these plants help to produce worldwide.

It is this problem—global air pollution—that presents the true strategic threat to which we must now respond. The political battles against local air pollution are the easiest to organize because the direct effect on human health can be seen vividly in the hazy, smog-choked skies and heard loudly in the hacking and coughing of the affected citizenry. The battles to control regional air pollution are more complex because the people who are most affected often live in a different, downwind region from the people most responsible for causing it. Still, this problem is finally being addressed, even as heated arguments continue over cause and effect.

However, the political struggle to control atmospheric pollution at the global level has barely begun. Every person on earth is part of the cause, which makes it difficult to organize an effective response. But every person on earth also stands to suffer the consequences, which makes an effective response essential and ought to make it possible to find one—once the global pattern is widely recognized.

One threshold that must be crossed before we can recognize the global pattern is the prevailing notion that the sky is limitless. Some of the pictures brought back from space by the astronauts and cosmonauts show that in fact the atmosphere is a very thin blue translucent blanket covering the planet itself. The earth's diameter is one thousand times greater than the width of the translucent atmosphere surrounding it; to put it in perspective, the distance from the ground to the top of the sky is no farther than an hour's cross-country run. The total volume of all the air in the world is actually quite small compared to the enormity of the earth, and we are filling it up, profoundly changing its makeup, every hour of every day, everywhere on earth.

> ***"It is this problem—global air pollution—that presents the true strategic threat to which we must now respond."***

We would prefer not to believe this, but consider the North Pole, far from any factory or freeway, where pollution known as Arctic Haze now reaches levels during the winter and spring that are comparable to the levels of pollution in many large industrial cities. Scientific analysis indicates that most of the Arctic Haze originates in northern Europe, making it, in effect, a particularly extensive example of regional pollution. Nevertheless, it illustrates the point that air pollution now reaches every place on earth. Air samples in Antarctica make the same point.

But the most troubling strategic air pollution threats are those that are truly ubiquitous and uniform throughout the world. Ironically, these threats are least likely to cause anyone immediate and direct personal harm, and consequently they are often perceived as benign. However, they are the changes most likely to do serious and lasting damage to the ecological balance of the earth itself.

The molecules of the air exist in a state of equilibrium; similarly, the atmosphere exists in a state of dynamic equilibrium with itself and with life on the planet. Dramatic changes in that equilibrium in just a few decades can threaten the balancing role played by the atmosphere within the larger global ecological system.

Most things on earth have adjusted through the eons to an amazingly persistent and stable balance in the makeup of the global atmosphere. The relatively small number of air molecules in the atmosphere have been continuously recycled through animals and plants since oxygen was first produced in large volume by photosynthetic microorganisms almost 3 billion years ago. Those animals and plants have adapted, over long periods, to the precise combination of molecules that have been present in the air throughout most of evolution, and they have, in turn, affected the composition of the atmosphere.

In every breath we take, we bathe our lungs in a homogeneous sample of that same air—many trillions of molecules of it—with at least a few in each breath that were also breathed by Buddha at some point during his life, and a like number that were breathed by Jesus, Moses, Mohammed—as well as Hitler, Stalin, and Genghis Khan. But the air we breathe is profoundly different than was theirs. For one thing, mixed in with the air molecules are a variety of pollutants, which vary according to where we live. More important, however, the concentration of some natural compounds has been artificially changed everywhere on earth. For example, every single person alive now inhales with each breath 600 percent more chlorine atoms than did Moses or Mohammed. The chemicals responsible for all this extra chlorine—now ubiquitous in the world's air—were first used in world commerce less than sixty years ago. That extra chlorine doesn't directly affect human health as far as we know, but it has a dangerous and debilitating strategic effect on the healthy functioning of the atmosphere. Like an acid, it burns a hole in the earth's protective ozone shield above Antarctica and depletes the ozone layer worldwide.

Ozone depletion is, in fact, the first of three strategic, as opposed to local or regional, air pollution threats; the other two are diminished oxidation of the atmosphere (a little-known but potentially serious threat) and global warming. All three have the power to change the makeup of the entire global atmosphere and, in the process, disrupt the atmosphere's crucial balancing role in the global ecological system. . . .

One of the clearest signs that our relationship to the global environment is in severe crisis is the floodtide of garbage spilling out of our cities and factories. What some have called the "throwaway society" has been based on the assumptions that endless resources will allow us to produce an endless supply of goods and that bottomless receptacles (i.e., landfills and ocean dumping sites) will allow us to dispose of an endless stream of waste. But now we are beginning to drown in that stream. Having relied for too long on the old strategy of "out of sight, out of mind," we are now running out of ways to dispose of our waste in a manner that keeps it out of either sight or mind.

> ***"Extra chlorine . . . has a dangerous and debilitating . . . effect on the healthy functioning of the atmosphere."***

In an earlier era, when the human population and the quantities of waste generated were much smaller and when highly toxic forms of waste were uncommon, it was possible to believe that the world's absorption of our waste meant that we need not think about it again. Now, however, all that has changed. Suddenly, we are disconcerted—even offended—when the huge quantities of waste we thought we had thrown away suddenly demand our attention as landfills overflow, incinerators foul the air, and neighboring communities and states attempt to dump their overflow problems on us. . . .

> ***"In the United States, the percentage of municipal waste incinerated more than doubled . . . in just four years."***

The latest scheme masquerading as a rational and responsible alternative to landfills is a nationwide—and worldwide—move to drastically increase the use of incineration. In the United States, the percentage of municipal waste incinerated more than doubled — from 7 percent in 1985 to over 15 percent in just four years—and investments in new incineration capacity are expected to double that percentage again in the next several years. In some of these projects, the heat generated by the incineration process is used as a source of energy to make steam, which is then sold to help offset the cost. In still other designs, the waste is molded into burnable pellets of "refuse-derived fuel." But even though the virtue of converting waste to energy is widely touted, the actual amount of energy produced is small and the principal and overwhelming reason for building such plants is that something has to be done with the massive amounts of garbage we create.

The huge new investment in new incinerators—almost $20 billion worth—is being made even though major health and environmental concerns have never been adequately answered. According to congressional investigators, the air pollution from waste incinerators typically includes dioxins, furans, and pollutants like arsenic, cadmium, chlorobenzenes, chlorophenols, chromium, cobalt, lead, mercury, PCBs [polychlorinated biphenyls], and sulfur dioxide. In the case of mercury emissions, a lengthy study by the Clean Water Fund found that "municipal waste incinerators are now the most rapidly growing source of mercury emissions to the atmosphere. Mercury emissions from incinerators [have] surpassed the industrial sector as a major source of atmospheric mercury [and] are likely to double over the next five years. If the incinerators under construction and planning come on line, with currently required control technology, mercury emissions from this source are likely to double. This growth will add millions of pounds of mercury to the ecosystem in the next few decades unless action is taken now." Mercury, of course, does not break down in the environment but rather accumulates, especially in the food chain, by means of a process called bio-accumulation, which concentrates progressively larger amounts in animals at the top of the food chain, such as the fish we catch in lakes and rivers.

The principal consequence of incineration is thus the transporting of the community's garbage—in gaseous form, through the air—to neighboring communities, across state lines, and, indeed, to the atmosphere of the entire globe, where it will linger for many years to come. In effect, we have discovered yet another group of powerless people upon whom we can dump the consequences of our own waste: those who live in the future and cannot hold us accountable.

Air Pollution Depletes Earth's Ozone Layer

by John Gribbin

About the author: *John Gribbin is a science writer who holds a Ph.D. in astrophysics from Cambridge University in England. Gribbin, a science topics contributor to the British Broadcasting Corporation and* New Scientist *magazine, is the author of* In Search of the Big Bang *and* The Hole in the Sky, *from which this viewpoint is excerpted.*

There have been dramatic new developments in our understanding of the holes in the ozone layer and in our scientific assessment of the spreading risk of ozone depletion. The most important development is that I can now refer to "holes" in the plural, for it is now clear that ozone depletion occurs over the north polar region, as well as above Antarctica, although as yet the depletion is less severe in the Northern Hemisphere. Going hand in hand with this discovery, there has been unambiguous confirmation that the cause of this ozone depletion is indeed the action of chlorine, originally from the chlorofluorocarbons (CFCs) released into the atmosphere by human activities, reacting with ozone in the stratosphere. Antarctic ozone depletion occurs *exactly* where there is active chlorine, in the form of chlorine monoxide, ClO.

Solar Influences

While these developments were taking place, the fact that the Antarctic hole itself was as big as ever at the end of the 1980s and in the early 1990s, at a time when the Sun was at a peak of activity, showed once and for all that the ozone depletion is not caused by changes in the Sun. Any solar influence on the ozone layer acts to make *more* ozone when the Sun is more active, and has therefore been partially canceling out the ozone depletion in the past few years. As solar activity declines up to the mid-1990s, the solar effect will also be acting to deplete ozone, adding to the depletions caused by CFCs and making the situation in both Northern and Southern Hemispheres worse.

All of that, however, pales into insignificance alongside the most dramatic new discoveries of the 1990s. First, it is now clear that the amount of ozone in the Northern Hemisphere stratosphere fell by 8 percent between 1979 and 1990, across a band of latitudes between 30N and 50N that covers virtually all of the contiguous United States, as well as Europe from the Mediterranean to the south of England. Then, the eruption of Mount Pinatubo, in the Philippines, focused renewed attention on the possibility that the debris from such an eruption, blasted upward into a stratosphere increasingly laden with chlorine from CFCs, could encourage further ozone depletion. All the evidence is that it does, and that Pinatubo itself is having a detrimental effect on stratospheric ozone.

"The worldwide depletion of stratospheric ozone . . . will get worse before it gets better."

So all the signs are that ozone depletion is intensifying as the 1990s progress. . . .

The new evidence is, indeed, so compelling that in response to the growing threat, the forty-nine original signatories to the Montreal Protocol have now called for a *total* ban on CFC production by the year 2000. But even if this is achieved, the atmosphere will still contain more CFCs in the year 2000 than it does today, and 90 percent of the CFCs in the air when the ban comes into force will still be there in 2010. The worldwide depletion of stratospheric ozone, not just at the poles but literally over our heads, will get worse before it gets better. . . .

Early Uses of CFCs

For decades, CFCs had been regarded by the chemical industry as something of a miracle substance. Their unusual properties had been discovered more or less by accident at the end of the 1920s, when they were originally developed as the working fluid used in refrigerators. They seemed ideally suited to the task, since they boiled at between –40 degrees F and 32 degrees F, were nonflammable, nontoxic, cheap to manufacture, easy to store, and chemically stable. It is indeed precisely because they are very stable compounds that CFCs do not react either with oxygen, to burn, or with living things, to poison them. They seemed perfect for use as propellants in spray cans, and the first product of this kind using CFCs went on the market in 1950. They have also proved useful as solvents, effective at cleaning delicate semiconductor circuitry without attacking the plastic boards on which the circuits and chips are mounted, and in blowing foams of all kinds, from fire extinguisher foam to the foam insulation used in the walls of some houses, and the foamed hard plastic of disposable coffee cups and clamshell hamburger cartons. . . .

CFCs had seemed too good to be true, and so it proved. A great deal of these chemicals were getting out into the environment and beginning to build up there. Seventy-five percent of the emissions to the air came from spray cans, which by their very nature had to release the propellant gases; about 15 percent came from leaky refrigeration and air-conditioning systems, especially car air conditioners. CFCs are so stable that they remain in the atmosphere for a very long time. There is nothing in the troposphere that can attack them and break down their chemical structure. They are so stable, indeed, that Jim Lovelock, the first scientist to investigate their distribution around the world, did so because he was interested in tracing the movement of air currents, and decided that CFCs would provide an excellent marker showing how air masses moved. . . .

The Danger of CFCs

The potential problem with CFCs is that they contain chlorine. Indeed, the full name for this class of compounds, chlorofluorocarbons, indicates that they are built up from atoms of chlorine, fluorine, and carbon. Du Pont, the major manufacturer of CFCs, gave their own products the brand name Freons, and developed a labeling system that indicates, to the cognoscenti, how many atoms of each kind are present in a particular type of Freon. This system is now widely used for all CFCs; a shorthand naming system has evolved in which the two most commonly used CFCs are known as F-11 and F-12—their full chemical names are trichlorofluoromethane (CCl_3F) and dichlorofluoromethane (CCl_2F_2). Others that are important to the debate today are F-22 ($CHClF_2$) and F-113 ($C_2Cl_3F_3$). F-11, it is now estimated, survives for 75 years before being broken down in the atmosphere, and F-12 for 110 years—so pollution being caused by release of these products now will still be affecting our planet at the end of the twenty-first century, even if CFC emissions are halted tomorrow. . . .

"[CFC] pollution . . . will still be affecting our planet at the end of the twenty-first century."

On the principle "know your enemy," the first thing to do is to check out just where CFCs are widely used today. Although the United States, Canada, and Sweden have banned most uses of CFCs in spray cans, in the mid-1980s this still represented about a third of the use of F-11 and F-12 worldwide, at least in the countries reporting to the Chemical Manufacturers Association, which represents some 85 percent of estimated global production. Together, the two CFCs still account for about 70 percent of all emissions, with a further 12 percent (and rapidly growing) coming from F-113, which is widely used as a solvent in the burgeoning microchip industries of countries like South Korea. F-12 is used in large quantities in the United States and Japan, in particular, as the working fluid in automobile air conditioners, and the main use of F-11 today is as a foam-blowing agent.

Quite astonishing amounts of F-12 are wasted by automobile air conditioners. In 1985, the United States produced 150,000 tons of F-12, and one-third of this went into such systems. Thirty percent of the fluid in these systems is lost by "routine" leakage, and half escapes during servicing. The rest is released when the units are eventually scrapped.

Both F-11 and F-12 are used in refrigerators, and ought to be securely sealed away there, posing no problem. But these units, too, are scrapped at the end of their useful lives (or when the owners get bored with them and decide to buy a new model), and then the CFCs are set free.

Although the bubbles in your Styrofoam cup or hamburger carton may contain F-11, the gas inside them is not a major threat to the ozone layer. Very little of it escapes. What is a problem is the amount of F-11 that escapes during the manufacturing process, when the plastic is foamed in the first place. This use of CFCs is rapidly expanding. Half the 83,000 tons of F-11 produced in the United States each year in the late 1980s went into rigid foam sheets and packages (along with 11 percent of the F-12 produced), and in Britain manufacturers of so-called expanded plastic predicted 10 percent growth in 1987.

Controls on CFCs

As these examples show, there is a huge potential to reduce the amount of CFCs from getting into the atmosphere simply by making existing uses more efficient, controlling the release from factories that make foamed plastics, sealing air conditioners more effectively, and recycling the CFCs when units are scrapped. Unfortunately, CFCs are so cheap to manufacture that there has been no incentive for any of these techniques to be developed in the past. One very obvious way to provide such an incentive would be by taxing the CFCs; this could have a dual benefit if money raised in this way were spent on developing replacement chemicals. . . .

In June 1990, progress seemed to be being made when delegates from more than 80 countries met in London to discuss ozone depletion and, it was hoped, beef up the Montreal agreement. In spite of sometimes-bitter arguments between the representatives of the developed world and those from poorer countries over who should pay for the replacement of CFCs by ozone-friendlier substitutes, by the end of that meeting one typical headline blazed "Deal to save ozone layer agreed" above a story reporting that even the representatives from India and China had said they would recommend their governments to sign the Montreal Protocol. Yet a full year later, in July 1991, a much more modest story buried in the

"There is a huge potential to reduce the amount of CFCs from getting into the atmosphere."

newspapers reported that although India had decided in principle to sign, a final decision was still awaiting news of whether the industrialized countries would pay the $1.2 billion it needed to make the switch, and that although China had signed the toughened Protocol in June 1991, it had not ratified the agreement. Under the deal reached in June 1990, which looked so promising at the time, the richer countries pledged a seemingly modest $240 million to a fund to aid the transfer of ozone-friendly technology to poorer countries; by July 1991, only $9.3 million had been paid into the fund.

A Lasting Problem

There is little chance that the richer countries will meet even this modest bill, which is itself far short of India's own stated requirement, and in the short term some of the substitutes for CFCs may do more harm than good to the ozone layer. Researchers continue to develop alternatives to CFCs, and to investigate the consequences of releasing those alternatives into the atmosphere. Refrigerators that work on a different principle (known as the Stirling cycle) from the ones you are familiar with, and which work without the use of CFCs, are among the many alternative products under investigation, and polystyrene manufacturers are quick to point out that their industry now accounts for less than 2 percent of all CFCs used in the U.S., switching to HCFC-22 [a less risky hydrochlorofluorocarbon] instead. But while every little bit helps, and any moves in this direction should be encouraged, it is still true that ozone depletion will get worse before it gets better. . . .

"CFCs and other ozone-depleting chemicals continue to pour into the atmosphere at about the same rate as in 1990."

Whatever you may hear about worthy intentions and promises by the politicians, CFCs and other ozone-depleting chemicals continue to pour into the atmosphere at about the same rate as in 1990. Even if the strengthened version of the Montreal Protocol, promising a phaseout of the worst offenders by signatory countries within the decade of the 1990s, as agreed at that London meeting in 1990, is strictly adhered to by all the nations (which is far from being the case yet), the burden of chlorine in the stratosphere will keep on growing until at least the year 2000, when it will be at least double the level at which the Antarctic ozone hole appeared. On the most *optimistic* scenario it will then slowly decline, but still be one-and-a-half times the level that triggered the appearance of the Antarctic hole in 2050. Speaking in 1990, [Antarctic hole discoverer] Joe Farman said that "we are condemned to 60 years of the unknown," and called on all nations to ban all CFCs and related products. Nothing less is good enough.

Air Pollution from Hazardous Waste Incinerators Endangers Public Health

by Peter Montague

About the author: *Peter Montague is the director of the Environmental Research Foundation, a public interest organization in Annapolis, Maryland.*

Hazardous waste incinerators, and the web of regulations intended to make them operate safely, have come under withering criticism from government scientists, private researchers and the *Wall Street Journal*. Officials of the U.S. Environmental Protection Agency (EPA) and private research scientists now admit that hazardous waste incinerators emit hundreds of times more dioxins and other toxic air pollutants than is allowed by EPA regulations, and the *Wall Street Journal* revealed a record of malfunctions, including explosions and major releases of toxins, that incinerator operators have tried to cover up and that regulatory officials seem powerless to understand, much less curtail.

Failing to Meet Regulations

Scientists employed by EPA acknowledged in March 1992 that modern hazardous waste incinerators simply cannot comply with existing federal regulations because they cannot destroy all chemicals with 99.99 percent destruction/removal efficiency (DRE), which is the level required by federal law. Federal law further requires that certain wastes of "special concern," such as dioxins, furans and PCBs [polychlorinated biphenyls], be destroyed with 99.9999 percent DRE. EPA scientists said in April 1992 that they have known since at least 1985 that hazardous waste incinerators could not meet any of

Peter Montague, "Torching the Environment," *Multinational Monitor*, May 1992. Reprinted with permission of *Rachel's Hazardous Waste News* and the Environmental Research Foundation, Annapolis, Maryland.

these regulatory requirements.

The story broke when Pat Costner, a chemist and research director for Greenpeace, published an independent analysis of dioxin emissions from the Jacksonville, Arkansas incinerator. The Jacksonville incinerator has begun burning 16.5 million pounds of herbicides (2,4,5-T and 2,4-D) left over from the Vietnam War. These wastes are known to be contaminated with total dioxins and furans at concentrations ranging from 3 to 40 parts per million (ppm).

> ***"Hazardous waste incinerators emit hundreds of times more dioxins and other toxic air pollutants than is allowed."***

Costner's analysis revealed that the Jacksonville incinerator was only achieving 99.96 percent destruction of the dioxins entering the incinerator, thus emitting 400 times more dioxin into the community than the law allows. An official with the Arkansas Department of Pollution Control and Ecology acknowledged in telephone interviews that Costner's calculations are correct. He also said the department had no intention of shutting down the incinerator despite its continuing emissions of dioxin directly into a residential community. He said the department did not know what the total dioxin emissions into the population of Jacksonville would be. But, he asserted, no matter what the total may be, it is safe.

The Jacksonville incinerator is a key demonstration project, established with the cooperation of former EPA Administrator William Reilly and Arkansas Governor Bill Clinton to show that dioxin-containing wastes can be incinerated in a residential neighborhood over the objections of the community. In a citywide referendum in March 1986, the people of Jacksonville voted two-to-one (1383 to 656) to stop the project, but government officials simply ignored the vote and have overridden all objections ever since. Costner's analysis clearly showed that residents of Jacksonville are being exposed to levels of dioxin contamination that exceed federal health and safety standards by a wide margin. This is the first systematic dioxin experiment on humans using a residential population. Previous dioxin exposures of humans have occurred during industrial accidents and in the industrial manufacture of chemical-biological warfare agents. Dioxin is now known to cause cancer in humans and to disrupt normal growth and development of fetuses and infants at low levels of exposure.

License to Burn

About 100 waste sites in the United States contain substantial quantities of dioxin, and the United States has stockpiles containing billions of pounds of chemical-biological warfare agents which the federal government wants to incinerate. If the Jacksonville dioxin experiment can be maintained despite ethical and public health objections, government agencies will be able to claim they have a green light to incinerate just about anything just about anywhere.

The Jacksonville experiment has brought to light information that could derail the entire U.S. incineration program, however. In preparing her analysis of dioxin exposure of the Jacksonville populace, Costner uncovered a government study showing that tests conducted in 1984 and 1985 by private researchers under contract to EPA revealed that hazardous waste incinerators cannot be expected to achieve 99.9999 percent destruction of wastes that occur in concentrations lower than 10,000 parts per million, and cannot be expected to achieve 99.99 percent destruction of wastes that occur in concentrations lower than 1000 ppm. EPA published the 1985 data in 1989.

Commercial Hazardous-Waste Incinerators

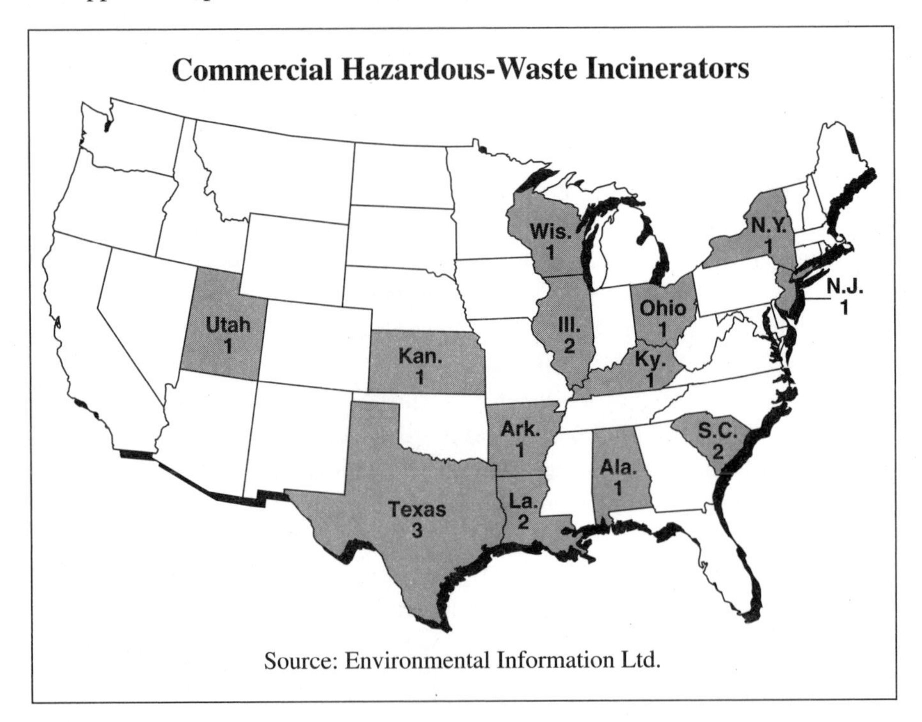

Source: Environmental Information Ltd.

When this information came to light, a news reporter from the *Arkansas Democrat-Gazette*, Sandy Davis, interviewed Bob Hall, chief of the EPA's Combustion Research Branch in Research Triangle, North Carolina, and he confirmed what the EPA report had shown. "The fact is that you run into problems with your DRE when a low concentration of wastes is fed into the incinerator," Hall said. "Our data clearly shows that." Davis asked Hall why EPA hasn't changed its regulations since it knows that existing incinerators cannot comply with the regulations. Hall said, "I don't know why that hasn't been changed. It's a regulatory issue. I'm in research."

Costner uncovered a second EPA report, published in 1984 but never widely circulated, showing that, among eight major hazardous waste incinerators studied, none could achieve 99.99 percent DRE. Sandy Davis interviewed the author of that report, Drew Trenholm of the Midwest Research Institute in Research Triangle, North Carolina, who said that incinerators simply cannot achieve the DRE required by federal law. "The trend is very strong in the data that this is the case," Trenholm told Davis.

At public hearings over the past decade, dozens of EPA officials have stated for the record that incinerators can achieve the legally required DREs in what appears to be a coverup of public health information of astonishing proportions.

Many of the most dangerous toxins, such as dioxins, furans and PCBs, occur in wastes at low concentrations. If low-concentration chemicals cannot be destroyed effectively, then all sludge incinerators, contaminated-soil burners and wood-treatment-waste incinerators will fail to meet federal regulations and will emit illegal quantities of potent toxins into surrounding air.

The Chem Waste Nightmare

The human management failures of incineration technology are no less remarkable than the failures of the technology itself.

Chemical Waste Management is the nation's largest and wealthiest operator of hazardous waste incinerators. Joan Bernstein, vice-president for environmental policy and ethical standards at Chem Waste, says, "Environmental compliance is what drives this company." Some of the company's parent firm's top executives donate time to sit on the boards of directors of prominent environmental organizations like the Audubon Society and the National Wildlife Federation. If any entity were capable of running an incinerator well, it would seem to be this company.

Yet during recent years Chem Waste's two incinerators have racked up a list of leaks, spills, releases, explosions, violations and coverups that would fill a hefty book.

According to *Wall Street Journal* reporter Jeff Bailey's review of Illinois state EPA inspection and other records and *Journal* interviews with state and company officials, among Chemical Waste Management's numerous violations of standard operating procedure are the following:

- Chem Waste has mixed incompatible wastes together, causing chemical reactions that sent plumes of waste wafting off-site.

> ***"[Jacksonville] is the first systematic dioxin experiment on humans using a residential population."***

- Chem Waste on several occasions has been caught feeding wastes to the furnace at excessive rates, reducing the effectiveness of waste destruction.
- Chem Waste sometimes burns wastes at temperatures of only 1300 degrees, far below the required 1600 degrees.
- Chem Waste has permitted carbon monoxide to exceed the limit of 500 ppm

in the stack gas.

• Chem Waste has failed to maintain proper manifests indicating where all the wastes come from.

• Chem Waste has failed to keep proper operating records.

• In some instances, Chem Waste has failed to transfer waste in leaking containers to new containers.

• All wastes are supposed to be sampled to avoid putting explosives into the furnace, yet both Chem Waste incinerators have exploded recently, offering clear evidence of failure to identify wastes properly.

• Chem Waste has failed to report explosions at its incinerators.

• Waste feed is supposed to cut off automatically when hydrochloric acid (HCl) in the stack gas exceeds 100 ppm, but Chem Waste had its cut-off set for 500 ppm HCl.

• Chem Waste's incinerators are not licensed to burn dioxins, but on December 3, 1991, Chem Waste's incinerator at Sauget, Illinois, burned a 25 milligram vial of dioxin—enough to provide a maximum allowable dose for 232,000 people.

Government Ineptitude

Government regulation has done little to curb industry abuses. The *Wall Street Journal*'s Bailey pointed out on March 20, 1992, that federal, state and local regulatory officials pay close attention to hazardous waste incinerators,but they can't be everywhere at the same time, and they often learn about accidents, explosions and violations from tips phoned to them anonymously by insiders. There are many other regulatory problems, as well.

"[There] appears to be a coverup of public health information of astonishing proportions."

The person responsible for developing EPA's hazardous waste incinerator regulations in 1978 was William Sanjour. In a recent letter to a grassroots activist, Sanjour offered several reasons why the regulations, as finally written, don't work:

"I've talked to many people who live near hazardous waste sites and I have reviewed many records, and this is the way it really works," Sanjour wrote. "Inspectors typically work from nine to five, Monday through Friday. So if the incinerator has anything particularly nasty to burn, it will do so at night or on weekends. When the complaints come in to the inspector's office the next day, he will call the incinerator operator and ask what's going on. He may also visit the plant, but he rarely finds anything. The enforcement officials tend to view the incinerator operator as their client and the public as a nuisance."

"Keep in mind that hazardous waste is a factory's garbage. If they typically ship out, say, 1,000 gallons a month of waste solvents and they find themselves with, say, 50 gallons of waste PCB which they don't know what to do with, what is more natural than dumping it in with the waste solvent to be hauled away to the incinerator? No one would be the wiser," Sanjour wrote.

He offered other reasons why the regulations are insufficient:

> ***"Both Chem Waste incinerators have exploded recently, offering clear evidence of failure to identify wastes properly."***

- The regulations require no monitoring of ambient air in the vicinity of the incinerator.
- It is easy for operators to cheat because they maintain the records.
- Government inspectors are typically poorly trained. They have low morale and high turnover. EPA statistics show that 41 percent of inspectors have conducted fewer than 10 inspections. "There is no reward to inspectors for finding serious violations and, indeed, zealous inspectors are typically given a hard time by their supervisors," Sanjour wrote.

Puffs of Pollution

Recent events at the Jacksonville, Arkansas, incinerator appear to follow a script that might have been written by Sanjour.

The Jacksonville site manager, Robert Apa, issued respirator masks to all employees and sent an interoffice memo in April 1992 ordering everyone to keep their masks handy because of dangerous "puffs" of pollution being emitted from the furnace. Seals in the fire box are leaking, and periodically, for reasons that are not understood, pressure builds up inside the furnace, forcing "puffs" of contaminants to escape. The puffs last from 5 to 45 seconds and represent emissions that entirely bypass the pollution control system.

When the media obtained an internal company memorandum discussing the puffs, Mark McCorkle, an Arkansas state official assigned to regulate the Jacksonville incinerator, first tried to pressure the *Arkansas Democrat-Gazette* (the state's largest paper) not to print anything about it. McCorkle then conceded that the pollution puffs posed a potential hazard to workers, but denied that the public would be affected when the puffs drifted off site to the homes that lie a long stone's-throw from the furnace. . . .

Jacksonville citizens continue to work desperately to shut down the incinerator. A new group, Jacksonville Mothers and Children Defense Fund (JAMAC), [planned to] file a lawsuit seeking a shutdown. They are asking groups across the United States to sign on to their suit.

After a decade of experimentation and experience, the record now indicates that hazardous waste incinerators cannot be operated safely, even when the operator desires to do so. If the operators have any inclination to cut corners, regulatory officials seem unwilling or unable to bring them to justice, intensifying the risks to the public even further.

Chlorine Destroys the Ozone Layer

by Adam Trombly

About the author: *Adam Trombly is a physicist and climatologist at the Institute for Advanced Studies, an Aspen, Colorado, research organization that studies environmental and geophysical phenomena.*

On the slopes in Aspen, we have had severe sunburn cases with people who were wearing Sun Protection Factor 15 suntan lotion out skiing for only two hours. Out in the oceans, coral reefs are being killed by ultraviolet radiation. On CNN [Earth Summit Secretary General] Maurice Strong admitted that ultraviolet radiation (UV) in South Africa, Tasmania and Chile is now 28 times the normal background level.

The sky isn't falling, it's tearing. I'm afraid Mama Gaia's not gonna take care of her babies, anymore. She's gonna kick her babies in the rear!

We've known there was an ozone hole over Antarctica since 1977 when the British Antarctic Expeditionary Group actually discovered it.

NASA and Political Pressure

The [reason the US couldn't see it] was because NASA [National Aeronautics and Space Administration] had written the program on the Nimbus satellite in such a way that anything *above* a three percent depletion of ozone wound up being recorded as "noise." After this story finally broke in 1985, NASA reevaluated its data and suddenly there *was* an ozone hole over Antarctica.

Since the Nimbus satellite makes 250,000 globe-scanning measurements an hour, the question arose: "What about the rest of the planet?" Following NASA's recalibration in 1985, the Nimbus also revealed a general, global ozone depletion. NASA still won't admit that it knew about this that far back. We've had global ozone depletion for years and it just kept getting worse. Now there have been warnings about 30-40 percent ozone loss over Helsinki, Toronto and Moscow.

Adam Trombly, "The Chlorine Cover-Up," *Earth Island Journal*, Fall 1992. Reprinted with permission of *Earth Island Journal*, 300 Broadway, Suite 28, San Francisco, CA 94133; (415) 788-3666. Earth Island Institute membership ($25/yr.) includes four issues of EIJ.

The National Center for Atmospheric Research (NCAR) scientists were being told what to say and what not to say. I have friends that I can't even talk to anymore. They're afraid to tell me what they know because they're afraid I'll tell the world. They don't want to lose their jobs.

It's politically inconvenient to acknowledge what the military-industrial complex has done to the atmosphere. In whose best interest is it that we should destroy our own planet? We're exterminating ourselves and we're all participating in the process through our passivity. . . .

Killer Chlorine

Chlorine is a chemical demon. One chlorine atom will destroy 100,000 ozone molecules. When Mario Molina and Sherwood Rowland predicted in 1974 that the ozone loss would reach 7-8 percent by the year 2050, it caused a panic. We reached that level in the 1980s. What accounts for the difference? Molina and Rowland were only figuring the chemical impacts for chlorine releases from CFCs. But this source only accounts for a small part of the destruction that has occurred. Where did the other chlorine come from?

The truth is that we have chlorine going up into the atmosphere from all kinds of sources. It's not just CFCs. It's chlorine from sewage treatment plants, water treatment plants, backyard swimming pools, even household chlorine bleach. We were told that these molecules were too heavy; that they couldn't get into the stratosphere, but it isn't true. Atmospheric scientists have known this for years—it's lunchtable conversation.

Not all chlorine rises into the air (some compounds become entrained in the soil or precipitate into water) but massive amounts of chlorine have been getting into the stratosphere the whole time.

Once the parent molecule—whether it's carbon-tetrachloride, polychlorinated hydrocarbon or CFC—gets into the stratosphere, it photo-reacts with the ultraviolet radiation and interferes with the production of ozone.

We've got to realize that all chlorine production must stop. This is the kind of talk that can get you killed. Chlorine is a huge industry. There is a tremendous amount of money involved. Do you realize the liability issue that is involved here? Remember asbestos? There was congressional testimony from the 1930s on the dangers of asbestos but nothing happened because the lobbies were too strong.

> ***"NASA reevaluated its data and suddenly there was an ozone hole over Antarctica."***

The first major extinctions in the world's oceans were caused by an abundance of oxygen. Anerobic organisms were killed off by the "aerobic revolution." Then came the invertebrates and the fish. In the atmosphere, membranes of ozone slowly developed, allowing the emergence of land life. This evolution of a natural sunscreen for the whole planet was a living process that took millions and millions of years to occur. And we've totally screwed it up in a couple of centuries.

The Montreal Convention [to control chloroflurocarbon gases] is an obscene compromise. We've compromised not only our lives but the lives of all species. The trees are dying and it's not from acid rain. In Colorado, only 15-20 percent of the soils were found to be acidic but 28 percent of the state's coniferous forests are now necrotic or dead. They've been sunburned by ultraviolet radiation.

> ***"Chlorine is a chemical demon. One chlorine atom will destroy 100,000 ozone molecules."***

We're being exterminated by congressional committees that turn their backs on the truth. When people figure out what's been going on, we're going to have War Crime Trials. Ten to twenty million people died at the hands of the Nazis, but now we're talking about *everybody* dying.

Why isn't anyone talking? These scientists are not evil people. They're trying to pay their mortgages. The economy's in a recession. We are working twice as much on a per capita basis as we were three decades ago and we're making a third less money. Have you tried to get a job as a physicist lately? Or a job as a climatologist—after you've been fired by NASA? You'll wind up pumping gas; you'll wait on tables.

The Kuwait Connection

The highest chlorine content of any oil in the world is found in Kuwait. The oil contains a lot of polychlorinated hydrocarbons and, from what I was told by the Petroleum Institute in Saudi Arabia, they have got quite a bit of free chlorine in those fields as well.

On June 8, 1991, NOAA [National Oceanic and Atmospheric Administration] said the top of the plume from the burning oil wells set aflame during the Gulf War was at 6800 feet. Look at the time of day they flew. They took off in the early morning [before the sunlight warmed the air and convention lifted the smoke skyward]. Flying at 1.23 PM in the afternoon, only 200 kilometers (124 miles) south of the source, we found that the top of the plume was at 18,000 feet.

This pollution did get into the stratosphere. In December 1991, at 56,000 feet, I saw all the way across the Atlantic Ocean, a brown layer of petroleum effluent. This also shows up in NASA STS 43 [space shuttle] photographs (which NASA still refuses to release to the public). The smoke from the fires was lofting all the way to Katmandu. A centimeter of goop fell on the ski slopes in Kashmir, according to Australia Broadcasting and the Brookings Institute. The smoke plume undoubtedly exacerbated the nucleation of the particulates in the Bangladesh monsoon that killed 135,000 people. Chemical condensation with oil in the atmosphere is highly capacitive, which is why we have more lightning strikes. And yet [EPA Administrator William] Reilly insisted that the fires were just "a regional problem."

We've got so much ultraviolet (UV) radiation coming in now, we've got secondary infrared heating from the reactions in the atmosphere. People need to get excited. We could have a second Jonestown worldwide. By our political indolence, we're committing suicide right now.

The '60s generation demanded and created the EPA. We got rid of Richard Nixon. Now we've got to demand an immediate halt to the use of chlorine. Boycott chlorine bleach. Demand that municipal water treatment plants stop using chlorine. True, you need to purify water, but you don't need to use chlorine—you can use ultraviolet radiation or ozonation at the inlets. It is done this way in science labs.

What Needs to Be Done

We need to begin atmospheric remediation programs on Manhattan Project levels worldwide. We need to supplement the amount of oxygen in the stratosphere immediately. In the labs we have found some remediation strategies that work pretty well, but it's not going to be easy. It'll cost billions of dollars—but it's an emergency!

Don't disband the army and add to the unemployment problem: we need an Environmental Defense Force. Use the military's bulldozers to reforest the deserts. Don't rely on these people who call themselves scientists: buy your own UV meter. The world environmental community has to try and set up a nonpoliticized research center—an independent International Center for Atmospheric and Forest Research. We need an International Environmental Constitution immediately—with big, big teeth—and we need people to be prosecuted.

Automobiles Cause Air Pollution

by Steve Nadis and James J. MacKenzie

About the authors: *Steve Nadis is a free-lance writer in Cambridge, Massachusetts. James J. MacKenzie is a policy analyst specializing in transportation issues at the World Resources Institute, an environmental research and policy institute in Washington, D.C.*

In 1989, ads for what appeared to be the environmentally perfect car began airing on TV. "The car is connected to the road," drones a sleepy New Age voice. The road, in turn, is connected to the Earth. The Earth, we learn, is one big ecosystem, which is part of the solar system, which is part of the Milky Way galaxy, *ad infinitum*. The car itself is called Infiniti.

Based on the early ads, the car seemed unique—a machine perfectly attuned to its natural surroundings. That's because there was no car. The pictures showed only a winding country road, trees rustling in the wind, water coursing in a bubbling brook, waves washing up on the seashore—all soft, soothing, and rhythmic.

Beyond the Hype

Months later, when Nissan finally unveiled the Infiniti, the company had to reveal the mythical chariot of its TV spots. That's when the magic ended. Behind the hype, we could see it was a car, just like other cars. It burned gasoline, gave off carbon monoxide, nitrogen oxides, hydrocarbons, and carbon dioxide, and it contributed to urban smog, rural air pollution, acid rain, and the buildup of greenhouse gases in the atmosphere.

There's nothing wrong with the Infiniti that's not wrong with most cars. Basically, an isolated car cruising down a quiet country lane is almost harmless. The problem is millions of cars on city streets and thoroughfares the whole world over. While the total number of motor vehicles—like the number of hamburgers sold by McDonald's—will never quite reach infinity, there has been a nearly eightfold increase since 1950. And the growth since then (at a rate of about 5.3 percent per year) shows no signs of tapering off.

The combined effects of the world's half-billion vehicles add up to a strikingly different picture than the pastoral vision presented in advertisements for the Infiniti. Instead of being in harmony with the environment, cars are among nature's chief adversaries. . . .

Something in the Air

Although astute observers may have sensed something funny in their air prior to 1950, that was the first time a scientific case was made linking cars to pollution in general and to the thick haze hovering over the Los Angeles basin in particular. In that year, Arie Jan Haagen-Smit, a chemist at the California Institute of Technology, proposed a theoretical mechanism for smog formation in which automobile exhaust and sunlight play central roles. His findings were hotly contested by the oil and auto industries. At a 1952 meeting of the American Petroleum Association, Vance Jenkins, research supervisor for the Union Oil Company of California, called the theory "unproved speculation" that could have dangerous repercussions for the business world. Within a few years, however, Haagen-Smit's "speculation" had earned virtually unanimous support among scientists.

> ***"Instead of being in harmony with the environment, cars are among nature's chief adversaries."***

In the four decades since Haagen-Smit first promulgated his theory, much has been learned about the automobile's contribution to dirty air. In the United States, motor vehicles contribute about 53 percent of the carbon monoxide, 30 percent of the nitrogen oxides, and 27 percent of the hydrocarbons emitted to the air. Ozone, the chief ingredient of smog, is produced when hydrocarbons from cars and other sources (including trees) and nitrogen oxides react in the presence of sunlight. (Ground-level ozone, which is a pollutant, should be distinguished from the natural ozone in the upper atmosphere, which screens Earth from harmful ultraviolet radiation.) Over the course of its lifetime, a typical new car equipped with pollution-control devices will spew out some 300 pounds of smog-forming compounds and 34 tons of carbon dioxide. Greater Los Angeles, home of Haagen-Smit's research, stands out in this respect, with tailpipes from eight million cars and trucks supplying 70 to 80 percent of the area's noxious fumes. Massachusetts Commissioner of Environmental Protection Daniel Greenbaum estimates likewise that 75 percent of his state's air pollution comes from cars and trucks.

Auto emissions and gasoline vapors contain a host of toxic and carcinogenic pollutants. For example, motor vehicles are responsible for about 85 percent of the benzene and 30 percent of the formaldehyde released to the air—chemicals that the Environmental Protection Agency [EPA] considers probable human carcinogens. Other carcinogenic vapors are given off when cars are on the road or at the gas station. Some brake linings contain asbestos—yet another source of potential carcinogens.

Despite these dire figures, American cars today are far cleaner than the pollution machines of the 1960s. After emission standards were first imposed in 1968, and then tightened further with the 1970 Clean Air Act, carbon monoxide and hydrocarbon emissions in new cars dropped by 96 percent, while nitrogen oxides emissions were slashed by 76 percent. These reductions came about largely through the adoption of the catalytic converter, which turns hazardous exhaust gases into less harmful ones. Engine modifications and the introduction of lead-free gasoline also helped.

Before the Clean Air Act was adopted in 1970, industry leaders lobbied furiously against the new emission limits, claiming that pollution reductions would be technically impossible to achieve as well as economically ruinous. "This bill is a threat to the entire American economy and to every person in America," Lee Iacocca, then vice president of Ford Motor Company, claimed in 1970. Despite these patriotic protestations, Detroit did help cut pollution dramatically and later trumpeted these improvements as evidence of the automakers' "can-do" record and their dedication to a cleaner environment. "In a way, we made liars out of ourselves because we sincerely believed we couldn't pull a rabbit out of the hat," Greg Terry, a GM spokesman, recently admitted.

Dismal Air Quality

Unfortunately, the impressive reductions in tailpipe emissions did not end pollution. In fact, air quality in many American cities has stayed the same or worsened. From 1987 to 1989, according to tests conducted by the Environmental Protection Agency, 119 U.S. cities exceeded federal standards for ozone or carbon monoxide, up from 64 cities that failed to comply between 1985 and 1987. Since 1986, anywhere from 30 to 50 percent of all Americans have lived in areas where federal air quality standards were violated at least occasionally. Much of this pollution is due to motor vehicles.

"Auto emissions and gasoline vapors contain a host of toxic and carcinogenic pollutants."

The Los Angeles region, not surprisingly, has the poorest record. There, air quality standards are exceeded an average of 137.5 days a year. On bad days, ozone levels in the city are three times the federal limit. 1988 was a particularly bad year; federal limits were breached on 232 days. Houston, the second worst city, violated standards on 30 days in 1988. Air pollution-related illnesses, according to estimates made by the American Lung

Association, cost the nation tens of billions of dollars per year in health-care expenses and lost productivity.

Of course the United States is not alone in its air pollution problems. Many other countries are worse off. In metropolitan Athens, for instance, the death rate jumps 500 percent on the most polluted days. Mexico City, with perhaps the world's dirtiest air, routinely exceeds ozone limits set by the World Health Organization by a factor of three or more. A record was established on March 17, 1992, when ozone reached four times the health standard. While emissions from the city's 36,000 factories and other airborne contaminants are also culprits, by far the biggest source of filth is the three million vehicles that clog the streets, burning five million gallons of low-quality fuel daily and emitting over 80 percent of the pollutants. "It appears that Mexico City exceeds our federal ozone standard every day of the year," claims former EPA official Michael Walsh.

"Air pollution-related illnesses . . . cost the nation tens of billions of dollars per year."

Scarce Pollution Control

Other cities—among them, Bangkok, New Delhi, and Santiago—face similarly immense air quality challenges. The situation in developing countries is comparable, at least in one respect, to that faced in the United States some three decades ago: the catalytic converter and other pollution control technologies are scarce where they exist at all.

A big part of the problem stems from a lack of money. Howard Applegate, an environmental engineer at the University of Texas at El Paso, has met repeatedly with officials at Mexico's environmental agency, SEDUE. They agree that vehicles are the principal source of the region's bad air, but add, "It all costs money, and we don't have it." A good tune-up, for example, can cut down on pollution significantly, Applegate notes, "but in Mexico, buying a set of points and a set of plugs for a six-cylinder car costs a week's wages."

In Mexico City, as in most major cities around the world, the problem is simple: *too many dirty cars*, which clutter the streets and cloud the skies. But the obvious solution—driving fewer cars, cleaner cars, fewer miles overall—is far from simple to bring about. Although American assembly lines began turning out lower-emission vehicles in the 1970s, resultant gains were partly offset by the number of cars on the road and the number of miles they travel—both of which have risen dramatically.

A growing auto population, for example, has kept New York City from meeting standards set for certain pollutants by the federal government. What's more, says Eric Goldstein of the Natural Resources Defense Council in New York, "if traffic keeps increasing, we may never achieve national health standards."

Further emission reductions can be achieved, to be sure. The Clean Air Act, signed into law by George Bush on November 15, 1990, calls for lowering the emission limits on hydrocarbons by 40 percent and on nitrogen oxides by 60 percent in all new cars by 1996. Beginning in 1992, oil companies are required to sell cleaner-burning gasoline in the cities where carbon monoxide pollution is worst. In March 1992, Bush announced that the EPA would not require automakers to install on-board canisters—conditionally required by the 1990 amendments to the Clean Air Act—to trap smog-forming compounds that would otherwise escape during refueling. EPA waived the canister requirement in favor of vapor collector systems to be installed at gas pumps. Finally, under a trial program, new "alternative-fuel" vehicles (including electric cars and vans) will gradually hit the streets in California. The California Air Resources Board will require that 2 percent of the new cars sold in the state by 1998 will have zero emissions. By 2002 that figure will rise to 5 percent, and by 2003 up to 10 percent. By 2010, according to an estimate made by the South Coast Air Quality Management District, 17 percent of passenger cars and light trucks in Los Angeles will be powered by electricity.

The auto industry fought tooth and nail against the 1990 bill, resurrecting many of the same arguments used two decades before. "We can't squeeze much more out of the catalytic converter," complained the device's inventor, Dick Klimisch, now General Motors' director of environmental activities. Former EPA official Michael Walsh has a different perspective. The industry's "public posture is always much more pessimistic than technical reality," he says. "When you give them a challenge, they meet it."

Winning the Battle

Truly coming to grips with motor vehicle air pollution means addressing two critical issues. First, clean air regulations pertain almost exclusively to new cars, but older, poorly maintained cars are the biggest offenders; 10 percent of the cars on the road spew out about half of the pollutants, the EPA claims. Inspection and maintenance programs aimed at older cars have not been as successful as the EPA had hoped.

"Older, poorly maintained cars . . . spew out about half of the pollutants."

Second, increased driving—the result of ever more vehicles being driven more—will probably offset much of the hard-earned cut in emissions from individual cars. Federal standards set limits on the amount of pollutants a car can release per mile, but not on the number of miles Americans can drive. Until we gain control over the number of vehicles on the road and the total miles they travel, we just can't win the battle against pollution.

Global Warming and Ozone Depletion Are Not Serious Problems

by Dixy Lee Ray

About the author: *Dixy Lee Ray is a former governor of Washington and former chair of the Atomic Energy Commission. Ray is a writer in Fox Island, Washington, and a member of the board of advisers for the American Council on Science and Health, a New York City association of doctors and scientists concerned with public health and with calming public fears concerning pollution and other environmental issues. She is co-author of the book* Environmental Overkill: Whatever Happened to Common Sense.

If you assume from the title of this article ["Are the Global Threats Real? Scientific Facts vs. Environmental Myths"] that I have a somewhat skeptical and irreverent attitude toward such popular environmental scenarios as "Global Warming" and "Ozone Depletion," you are correct. Yet it appears that nearly everyone believes that these are important problems from which the Earth must be saved! Why? Well, because everyone says so. But what of the evidence? What are the data that support these issues? Are there any contrary facts?

Global Warming

The claim is that the earth is warming up and that human activity—burning fossil fuels which increases the CO_2 [carbon dioxide] content of the atmosphere—is the cause. Moreover, the consequences of global heating are claimed to be disastrous including changes in weather, agricultural crops, sea level, *etc.*

Before examining the evidence, let's look back at a similar set of claims made less than 20 years ago. The issue was not global warming but global cooling! It was widely believed by most climatologists at the time that a new ice age was imminent. Here's what they said:

Dixy Lee Ray, "Are the Global Threats Real?" *Priorities*, Spring 1992.

> There are ominous signs that the earth's weather patterns have begun to change dramatically and that these changes may portend a drastic decline in food production—with serious political implications for just about every nation on earth. (Peter Gwynne, *Newsweek*, April 28, 1975)

> According to the Academy [National Academy of Sciences] report on climate, we may be approaching the end of a major interglacial cycle, with the approach of a full-blown 10,000-year ice age a real possibility . . . with ice packs building up relatively quickly from local snowfall that ceases to melt from winter to winter. (*Science*, March 1, 1975)

> The continued rapid cooling of earth since World War II is also in accord with the increased global air pollution associated with industrialization, mechanization, urbanization, and an exploding population, added to a renewal of volcanic activity. . . (Reid Bryson, "Environmental Roulette," *Global Ecology: Readings Toward a Rational Strategy for Man*, John P. Holdren and Paul R. Ehrlich, eds., 1971).

How similar these warnings sound to what is said today about global warming! Are our memories so short? Are the scare stories about climate as serious and as frightening as the activists in these areas would have us believe? I think not.

Predictions of Catastrophe

For more than twenty years the American public has been subjected to a barrage of criticism about the way we live, what we eat, what we manufacture, how much and what kind of energy we use and how we handle the inevitable waste products of our activities. Most recently we are told that we are destroying the earth and its capacity to support life. These scoldings include predictions of catastrophe unless we make fundamental, unpleasant and costly changes in the way we live. They have become a virtual litany of impending disaster, a crusade to "Save the Planet." The charges are very serious; the question is, are they right?

The Earth and its atmosphere constitute a "greenhouse." If that were not the case, our planet would respond to the sun's radiation the same as does the moon whose temperature during the lunar day may reach 212°F and drop to –270°F during the lunar night. On Earth, roughly 30 percent of the incoming solar radiation is reflected back into space by the atmosphere, 20 percent is absorbed and 50 percent reaches the Earth's surface to cause warming. Of this 50 percent, some fraction is reflected back as infrared radiation which in turn may be absorbed by certain constituents of air, the so-called "greenhouse gases" (carbon dioxide, methane, hydrocarbons, and water vapor). Any increase in the relative amount of these

> ***"It was widely believed by most climatologists [less than twenty years ago] that a new ice age was imminent."***

gases will, theoretically, result in elevated air temperatures.

The theory is well established and supported by both theoretical and experimental evidence. There's only one problem—the theory doesn't match what has actually occurred in nature. If it did, the Earth would have warmed 2 to 4°C over the past 100 years. It has not. At best, there might have been about 0.5°C increase in temperature, but mostly that took place before 1940. There is other conflicting evidence:

• Temperature records taken in the northern hemisphere over the past century show no upward trend.

• Analysis of 135 years of surface ocean temperatures taken by ships at sea shows no upward trend.

• Analysis of temperature measurements by satellite from 1978 to 1988 (TIROS II) taken continuously, day and night over land and sea show no consistent change.

• Analysis of certain plant species in the U.S. shows an interesting picture. Citrus fruit used to grow in the Southeast as far north as the Carolinas. Now oranges will not ripen north of Orlando, FL. Florida had 24 severe killing frosts in the last 30 years, but only six in the 50 years prior. In 1990 the U.S. Department of Agriculture put out its first revised hardiness report for commercial crops since 1965. Temperature data from 14,500 measuring stations show that the area where crops can be grown without certain danger of a killing frost has moved 100 miles south in the last 50 years.

> ***"We are told that we are destroying the earth and its capacity to support life."***

A History of Climate Change

Such data should come as no surprise. Earth has a history of weather and climate change. In the last 10 million years, 17 ice ages have each lasted thousands of years followed by an abrupt warming with glacial retreat and a period of moderate temperatures in the northern hemisphere that lasted from 10,000 to 12,000 years. (It has been about 11,000 years since the end of the last ice age. So from a purely statistical basis and assuming that the earth continues these cycles of temperature change, we are indeed due for another ice age!)

We should recall that ice ages are not really a global phenomenon, but are characteristic of the northern hemisphere. During ice ages past, continental ice sheets did not form in South America, Africa, Southeast Asia or Australia. The temperatures in the tropics remained relatively unchanged.

Moreover, during the current interglacial, there have been significant climate shifts in the northern hemisphere. Recall the medieval "little optimum" (900 to 1100 AD). The Vikings sailed across an iceberg-free north Atlantic Ocean, settled Greenland and probably Labrador. This was followed by the "little ice age"

(1430-1850). Cold was then so intense that trees froze and exploded from internal ice buildup in southern England, and the Thames river froze solid in London (1814).

> ***"Temperature records taken in the northern hemisphere over the past century show no upward trend."***

What do we know about the atmospheric concentration of CO_2? Quite a bit, and yet, not enough. We know with considerable certainty that the CO_2 concentration in air has increased roughly 25 percent—from 280 parts per million (ppm) to 365 ppm—since the beginning of the industrial age. It seems easy to trace that increase to modern man's burning of fossil fuels, and that is what most people do. But the situation is not so simple. Consider:

• Prehistoric CO_2 levels also changed—100 million years ago there were not 350 ppm, but 3,000 to 5,000 ppm! This was obviously not due to industry.

• Humans pump an estimated seven billion tons of CO_2 into the atmosphere every year. Nature produces in the same period about 200 billion tons.

• Freeman Dyson of the Princeton Institute for Advanced Study has examined the sources and sinks for CO_2 and concludes that 50 percent cannot be accounted for. This corroborates previous conclusions derived by oceanographers.

• Plants love CO_2. A doubling of CO_2 content under controlled conditions results in a 30 percent increase in growth and yield. It also results in a plant that has stronger, larger leaves and stems and is more resistant to drought and disease.

From all the above we can conclude that both the temperature regime and the CO_2 picture deserve greater study and understanding before trillions of dollars are spent to mitigate a problem that may not exist, or if it does, may not be very important.

Stratospheric Ozone

Yet the supporters of the global warming theory are adamant. Dr. Stephen Schneider of the National Center for Atmospheric Research says:

> We need to get some broad-based support, to capture the public's imagination. That, of course, entails getting loads of media coverage. So we have to offer up scary scenarios, make simplified, dramatic statements, and make little mention of any doubts we may have. Each of us has to decide what is the right balance between being effective and being honest. (*Discovery*, October 1989, p. 47)

Despite all the press release hype about the ozone "hole," consider the following facts.

• The ozone layer is not stable; it is in a state of constant turbulence. Variations in its thickness occur on a seasonal basis and vary according to latitude. Annual fluctuations are up to 25 percent. Greater thinning (up to about 50 percent) can occur at the south pole. Significant thinning takes place at both poles but is greater in the Antarctic.

• Incoming radiation from the sun, specifically the UV spectrum, both creates and destroys ozone.

• The so-called "hole" or thinning is characterized not by a total loss of ozone but by a 50 percent depletion which appears annually. It is not permanent, but lasts about three to five weeks and the ozone is then reconstituted.

• There is no overall loss of ozone.

• Polar thinning is related to the polar vortex, a cyclonic storm that forms each year at the end of the Antarctic winter.

• Besides extreme cold for several weeks and return of sunlight, ozone "depletion" appears to require the presence of the chloride ion (Cl). The belief persists that the Cl comes from the chlorofluorocarbons (CFCs), mainly freons, but there is no documented proof of this. Chloride is one of nature's most abundant ions, with volcanic eruptions and oceanic storms as major sources.

So how much chloride comes from CFCs? About 0.75 million tons annually. Yet the amount of chloride calculated to be in the stratosphere at any one time is 50 to 60 times this figure.

If indeed chloride is necessary to the stratospheric breakdown of ozone, what is its source? There is no documented evidence of CFC molecules in the stratosphere. There are no measurements, only theory. Perhaps recently launched instruments to measure the composition of the ozone layer will remedy this situation.

"Prehistoric CO_2 levels also changed—100 million years ago there were not 350 ppm [parts per million], but 3,000 to 5,000 ppm!"

The so-called ozone "hole" was discovered in 1956 by the Cambridge meteorologist, Gordon Dobson, who devised the instrumentation and techniques of measuring stratospheric ozone. He considered the Antarctic ozone thinning to be an anomaly until the phenomenon occurred again in 1957 when he reported it as a natural annual event. The French investigators P. Rigaue and B. LeRoy also measured the "hole" in 1958 when it was thinner than at any time since. This was 30 years ago, before the widespread use of CFCs. Their conclusion was briefly stated, "The thinning is related to the polar vortex . . . and the recovery is sharp and complete."

Ultraviolet Radiation

Concern about the loss of stratospheric ozone relates to penetration of UV radiation. The thinner the ozone shield, the greater the UV penetration to the earth's surface. But, measuring instruments set up in the U.S. in 1974 show no increase in surface UV radiation.

Longterm UV radiation can cause skin cancer; this is well established. But people have been unduly frightened by not being told that there is more than one kind of skin cancer. Types related to too much UV (from sunbathing or tanning salons), are unsightly, irritating and annoying but curable in 99 percent of the cases. The more rare form is malignant melanoma. This is not related to exposure to UV, is frequently fatal and may be genetically determined. To imply that ozone loss (even if it occurred) would lead to an increase in malignant melanoma is a false and malicious misuse of science.

On April 4, 1991, EPA Administrator William Reilly said, "the ozone has thinned four to five percent. . . . This means 200,000 more cancer deaths over the next 50 years." He called the situation "grim." His statement is wrong, both as to the purported thinning and the skin cancers. Even if he were right, a five percent increase in UV exposure is far less than would be received by moving to a lower latitude. For example, UV exposure increases about 22 percent from Washington, DC to southern Florida. A journey from either pole to the equator subjects a person to a natural increase in UV radiation of 5,000 percent. Epidemiologic studies in Europe have shown the rate of malignant melanoma to be six times higher in the North than in the sunny Mediterranean region. If melanoma were the result of UV exposure, those rates should be reversed.

Banning CFCs

The Bush administration sought to speed the elimination of CFCs in response to a NASA [National Aeronautics and Space Administration] press release which stated there was an ozone hole over North America. No data were released to back up the statement which was based on a single set of measurements taken on January 20, 1992. It is important to note that no trend can be established by a single measurement or short term measurements. Further, no mention was made of the recent addition of Cl to the atmosphere by Mount Pinatubo in the Philippines which likely contributed to this phenomenon.

Those who would ban the production and use of CFCs on the basis of computer simulations and undocumented theory, choose to overlook the reasons why CFCs were developed and put into use in the first place. They are nonvolatile, nontoxic and present no direct hazards to living organisms. CFCs are used in refrigeration and air conditioning equipment, in fire fighting (halon foams) and in degreasing and cleaning electronic components. Despite many promises to the contrary, no substitutes have been developed and put into production. All of the proposed substitutes have turned out to be toxic, flammable, corrosive and inefficient. Use of any of them, or return to cumbersome, ineffective refrigerants like ammonia or sulfur dioxide would require a total redesign of equipment. In the U.S. alone there are 5,000 companies that use CFCs; the value of

> ***"There is no documented evidence of CFC molecules in the stratosphere. There are no measurements, only theory."***

the goods they produce is $28 billion per year. There are millions of individual and commercial refrigerating and air conditioning units. The capital investment exceeds $150 billion. The entire food transportation and marketing system throughout the western world depends upon refrigeration. Is it sensible to throw all this away on the flimsy evidence so far offered as a reason to ban CFCs? Why not simply seal the units better and recycle the freon? Politicized science is bad science.

Enormous Costs

In conclusion, careful scrutiny shows that evidence supporting both global warming and ozone depletion is sparse and questionable. Yet the U.S. has already entered into an international agreement to ban the production of CFCs—and the cost of freon has already gone up 30 percent. Dr. Richard Benedict, who negotiated the CFC ban on behalf of the U.S. has acknowledged that this action sounded the "death knell" for an important part of the chemical industry. Yet he insists the ban was necessary even though the scientific basis for it has not been established.

I believe that we are entitled to ask, "Why?" The costs are enormous, yet they pale by comparison with the financial burden put upon the American people if the "global warming" advocates prevail. . . . The [1992 Earth Summit] conference proposed to reduce the emissions of carbon dioxide to 1988 levels and to bring about a further 25 percent reduction by the year 2000. This cannot be accomplished without serious curtailment of industry and severe reduction in our standard of living. Estimates place the cost at more than $3 trillion. Does our firm knowledge of the problem and its possible consequences justify such a sacrifice?

My answer is NO.

Scientists Exaggerate the Extent of Ozone Depletion

by Ronald Bailey

About the author: *Ronald Bailey is an environmental journalism fellow at the Competitive Enterprise Institute, a Washington, D.C., think tank that promotes free-market environmental solutions. He is the author of* Eco-Scam: The False Prophets of Ecological Apocalypse, *from which this viewpoint is excerpted.*

A full-blown "ozone hole" rivaling the one that appears over Antarctica might open up over the United States, zapping Americans with damaging ultraviolet sunlight during the spring, warned NASA [National Aeronautics and Space Administration] scientists at an ominous early February 1992 press conference. *Time* showcased the story on the front cover of its February 17 issue, warning that "danger is shining through the sky. . . . No longer is the threat just to our future; the threat is here and now." Then-Senator Albert Gore thundered, "We have to tell our children that they must redefine their relationship to the sky, and they must begin to think of the sky as a threatening part of their environment."

Spooked by NASA, the Senate hastily passed by 96 to 0 an amendment demanding that President George Bush order the chemicals implicated in ozone destruction be phased out earlier than scheduled. Stung by the vote, Bush rushed the ban of the refrigerants known as chlorofluorocarbons (CFCs) forward from the year 2000 to the end of 1995.

NASA Jumps the Gun

Although NASA did not acknowledge it, the "danger" of an ozone hole opening over the Northern Hemisphere had already passed in less than a month after the existence of the putative crisis was announced. By late February, satellite data showed that the levels of ozone-destroying chlorine monoxide had dropped significantly and provided absolutely no evidence of a developing ozone hole over the United States. NASA waited until April 30, 1992, to announce at a press conference that a "large arctic ozone depletion" had been "averted." In other words, no ozone hole had opened up over the United States. *Time*, far from featuring the story on its cover, buried NASA's admission that there was no ozone hole over the Northern Hemisphere in four lines of text in its May 11 issue.

"Satellite data showed . . . absolutely no evidence of a developing ozone hole over the United States."

One NASA atmospheric scientist averred that his agency "really jumped the gun," while another drily commented that "it was perhaps premature for NASA to say that something drastic was about to occur." What was the rush? Why did NASA bureaucrats and scientists feel that they needed to frighten the American public?

The NASA revelations were exquisitely timed to bolster the agency's budget requests for its global climate change program, whose funding is slated to double by fiscal 1993. One NASA atmospheric scientist even wondered if it was only a coincidence that Senator Gore's book of apocalyptic environmentalism, *Earth in the Balance*, was published just days before NASA held its ozone press conference. After all, Gore chaired the subcommittee on Science, Space and Technology, which oversees NASA's budget.

A Question of Funding

"What you have to understand is that this is about money," Melvyn Shapiro, the chief of meteorological research at a National Oceanic and Atmospheric Administration [NOAA] laboratory in Boulder, Colorado, told *Insight* magazine. He added, "If there were no dollars attached to this game, you'd see it played on intellect and integrity. When you say the ozone threat is a scam, you're not only attacking people's scientific integrity, you're going after their pocketbook as well. It's money, purely money." Shortly after Shapiro's frank comments appeared, he was muzzled. He stopped talking to the press and told colleagues that he had been told to shut up by his superiors.

And NASA had another reason for jumping the gun. Environmental activists and their sympathizers in Congress and the bureaucracy were anxious to push President Bush into attending the big United Nations "Earth Summit" in June 1992. Senator Gore likened the alleged ozone crisis to global warming and urged the president to sign the global climate change treaty that was the centerpiece of the "Earth Summit."

By now everyone (94 percent of Americans according to one poll) has heard that the earth's protective ozone shield is wearing thin and even has a hole in it over the South Pole. The looming ozone catastrophe will purportedly bring humanity withered crops, collapsing terrestrial and marine ecosystems, skin cancer epidemics, and populations whose immune systems have been seriously compromised. The culprits in this drama are a group of industrial chemicals purveyed by greedy corporations to pampered and spoiled consumers. Ozone depletion is the perfect ecological morality play.

An Ozone Nuisance

In a morality play, unfortunately, there is no place for ambiguity. Yet the impact of man-made chlorofluorocarbons (CFCs) on the ozone layer is a complex question that turns on murky evidence, tentative conclusions, conflicting interpretations, and changing predictions. It is tempting to ignore these complications, abandon critical thinking, and join in the apocalyptics' call for *drastic action now*. But humanity would do so only in defiance of reality, for it turns out that ozone depletion, like other environmental dooms, is less a crisis than a nuisance, one that can and should be dealt with in a calm, deliberate, and scientific way.

"It's terrifying," exclaimed an overwrought John Lynch. The program manager for polar aeronomy at the National Science Foundation added, "If these ozone holes keep growing like this, they'll eventually eat the world." As usual [environmentalist] Paul Ehrlich is also alarmed. He warns, "Pure luck may have saved civilization from a catastrophic threat closely related to global warming—ozone depletion." Ehrlich calls ozone depletion "our worst near miss so far."...

One of the most effective scare tactics deployed by the apocalyptics during the ozone controversy was to predict massive increases in skin cancer as a consequence of reduced ozone. Most of us are familiar with the damage that UV [ultraviolet] light can cause through sunburns. The incidence of nonmelanoma skin cancer is strongly correlated with exposure to UV. The U.S. Environmental Protection Agency (EPA) predicts that for every 1 percent reduction in the ozone layer there will be a 3 percent increase in nonmelanoma skin cancers. The fact is that the incidence of skin cancer increases by 1 percent every 12 to 18 miles closer to the equator or every 150 feet higher up a person lives.

How trustworthy are the EPA's calculations? Not very, according to Temple University dermatologist Dr. Frederick Urbach, who is a consultant on the U.N. Environmental Assessment of ozone reductions. "You can crunch numbers in a computer and get whatever results you want to come out," says Urbach, laughing. He concedes that skin cancer rates have been going up dramatically in recent decades, but adds that "the increases are due to people spending more time outside, not more UV." The death rate for nonmelanoma skin cancer is negligible, less than 1 percent. Dr. Urbach notes, "It takes real talent for someone to die of nonmelanoma skin cancer. You basically have to ignore a hole in your skin for years."...

> ***"Ozone depletion, like other environmental dooms, is less a crisis than a nuisance."***

So why the furor over a possible ozone hole in the Northern Hemisphere in the spring of 1992? The chief reason was that atmospheric scientists detected elevated levels (1.5 parts per billion) of ozone-destroying chlorine monoxide. Despite the crisis atmosphere generated by NASA's publicity campaign in February 1992, scientists had been predicting since the summer before that global ozone might decline substantially in 1992. Why?

In June 1991, the Mount Pinatubo volcano in the Philippines blasted up to twenty million tons of sulfur into the skies. In the atmosphere, volcanic sulfur is transformed into sulfuric acid droplets, which act like polar stratospheric clouds by sequestering the nitrogen compounds that inhibit the formation of chlorine monoxide. The evidence strongly suggests that the 1992 chlorine monoxide peak in the Northern Hemisphere resulted from the sulfurous Pinatubo eruption. Linwood Callis, a scientist in the Atmospheric Sciences Division at NASA's Langley Research Center, found that after the El Chichón volcano erupted in the early 1980s, ozone was significantly reduced worldwide. In August 1991, Guy Brasseur predicted "a substantial ozone decrease especially in the mid- and high-latitudes, and especially in winter." He predicted wintertime ozone losses due to the volcano of up to 15 percent. NOAA's renowned atmospheric chemist Susan Solomon predicted even more dramatic reductions might occur, perhaps as much as 30 percent in the spring in the mid-latitudes. So when NASA created a Northern Hemisphere ozone hole scare, it wasn't exactly a secret that the ozone layer might become a bit frayed in 1992.

> ***"The death rate for nonmelanoma skin cancer is negligible, less than 1 percent."***

Dubious Warnings

"I couldn't understand why NASA didn't come out and say that this could be a very unusual year because of the volcanic eruptions, that maybe what we're seeing is something we'll never see again," mused David Hofmann, an ozone expert at NOAA. "Instead, they seemed to imply that this is the start of something really big. That really wasn't very wise. If there's a major ozone depletion seen this year, it's quite likely that it is related to the volcano." Of course, Hofmann failed to factor in NASA's budget considerations.

Despite elevated levels of chlorine monoxide and the attendant NASA hype, scientists found no evidence of an ozone hole opening up over the Northern Hemisphere in 1992. However, sulfur from Mount Pinatubo and some Southern Hemisphere volcanoes did hasten the opening of the annual Antarctic ozone hole in 1992. . . .

There is ample reason to doubt similarly catastrophic warnings about CFCs and the ozone layer. Climate catastrophists like Stephen Schneider and Carl Sagan claimed that CFCs were particularly potent greenhouse gases, adding as much as 25 percent to increased global temperatures. But they failed to take into account the first law of ecology: Everything is connected to everything else. Ozone, too, is a potent greenhouse gas and so when CFCs destroy it the atmosphere tends to cool. According to NASA, ozone decreases largely offset predicted increases in global temperatures due to CFCs. "What had been thought was a major greenhouse gas turns out to be having a cooling effect," noted EPA Administrator William Reilly.

Not a Global Emergency

Nevertheless, despite a great deal of continuing scientific uncertainty, it appears that CFCs do contribute to the creation of the Antarctic ozone hole and perhaps to a tiny amount of global ozone depletion. If CFCs were allowed to build up in the atmosphere during the next century, ozone depletion might eventually entail significant costs. More ultraviolet light reaching the surface would require adaptation—switching to new crop varieties, for example—and it might boost the incidence of nonfatal skin cancer. In light of these costs, it makes sense to phase out the use of CFCs.

But ozone depletion is certainly not the "global emergency" that environmentalists like Friends of the Earth's Elizabeth Cook say it is. Ozone depletion is not part of some generalized, overarching environmental crisis. It is a nuisance caused by specific chemicals, which are even now being replaced.

Exaggerated Fears

The normal processes of science and democratic decision-making have proved adequate to correct what might have become a significant problem. In 1990 our national and international institutions hammered out an agreement to control CFCs, the London Agreement to the Montreal Protocol, which takes the interests of all affected groups into account (though imperfectly). Calls to abandon a moderate course of action and push up the deadline for the CFC ban are based on exaggerated fears and unrealistic predictions. On the evidence so far, despite the lurid crisis-mongering of radical environmentalists, waiting for more information on CFCs and ozone did not cause any great harm to people or to earth's ecosystems, nor will it.

"There is ample reason to doubt . . . catastrophic warnings about CFCs and the ozone layer."

Radical environmentalists, like Ehrlich and Michael Oppenheimer of the Environmental Defense Fund, argue that the experience with ozone depletion should teach us to respond swiftly and dramatically to the threat of global warming. Rafe Pomerance of the World Resources Institute says the international negotiations over CFCs were merely a dress rehearsal for drastic reductions in carbon dioxide emissions aimed at preventing global climate change.

While replacing CFCs eventually will cost billions, the price tag for abating carbon dioxide could run as high as $600 billion *a year*.

The environmentalists are right to suggest that the example of ozone depletion is relevant to the debate over global warming. But the example bolsters the conclusion that humanity should be highly skeptical of environmental apocalypses. The relevant lesson is not "He who hesitates is lost," but rather, "Look before you leap."

Air Pollution Does Not Cause Global Warming

by Jocelyn Tomkin

About the author: *Jocelyn Tomkin is an astronomer at the University of Texas in Austin.*

In his best-selling book, *Earth in the Balance*, Vice President Al Gore speculates that global warming caused by the greenhouse effect will throw "the whole global climate system . . . out of whack," dramatically reducing rainfall in parts of the world already troubled by drought, melting the polar icecaps, raising ocean levels, and devastating low-lying countries such as Bangladesh, India, Pakistan, Egypt, Indonesia, Thailand, and China. "In the lifetimes of people now living, we may experience a 'year without winter,'" Gore writes. "We are carelessly initiating climate changes that could well last for hundreds or even thousands of years."

Mechanics of the Greenhouse Effect

Do such predictions have a basis in reality? Central to this question is the greenhouse effect. Contrary to the impression given by the mainstream media, the greenhouse effect is not of recent origin; it has been around for billions of years. But the link between this indisputable phenomenon and Gore's doomsday scenarios is tenuous at best. To understand why, you have to know something about the mechanics of the greenhouse effect.

The sun bathes the earth in sunshine. Some of the sunshine is reflected straight back into space, either by clouds or by the earth's surface. The remainder is absorbed by the earth's surface and thus heats it. Predictions about global warming hinge on the question of where this heat goes. Answering this question requires a brief trip into the world of electromagnetic radiation—light in all its forms, both visible and invisible.

All bodies radiate heat. A simple physical law says that the cooler a body is,

Jocelyn Tomkin, "Hot Air." Reprinted, with permission, from the March 1993 issue of *Reason* magazine.

the longer the wavelength at which it radiates its heat. The hot plate of a stove, for example, glows an orange-red when it's running at full blast. But turn it down, and as it cools it radiates its heat at longer and longer wavelengths, until it cools to the point where it's radiating exclusively in the infrared. To the eye it now appears to be off, although it is still too hot to touch.

> ***"The greenhouse effect is not of recent origin; it has been around for billions of years."***

Even everyday objects at everyday temperatures are busy radiators of heat in the infrared. A block of ice at melting point, for instance, has a temperature of 273 degrees Kelvin and is a raging furnace compared to a block of ice at absolute zero. (The Kelvin temperature scale is the same as the Celsius scale, except its zero is absolute zero, instead of the freezing point of water.) The earth itself continually radiates heat from both its dayside and its nightside. The balance between sunshine's heating effect and the cooling effect of the radiation the earth pours back into space allows the planet's surface to maintain a roughly constant temperature (apart from diurnal and seasonal variations).

But the earth's surface, with an average temperature of 288 degrees Kelvin, is much cooler than the sun's, with a temperature of 5,800 degrees Kelvin. So while the sun pumps most of its heat into space in the form of user-friendly visible light, the earth returns this heat to space in the form of much-longer-wavelength, invisible, infrared radiation. A greenhouse gas is a gas that is transparent at visible wavelengths but opaque at infrared wavelengths. It thus admits sunshine but blocks the escape of the earth's infrared radiation, thereby warming the planet's surface. As a rule, gases whose molecules have three or more atoms, such as carbon dioxide, are greenhouse gases, while gases whose molecules have only two atoms, such as oxygen, are not.

Carbon Dioxide: A Minor Player

Among the greenhouse gases, carbon dioxide gets the lion's share of attention because its concentration is increasing, largely due to industrial activity. But it is actually a minor player. If the concentration of carbon dioxide in the atmosphere doubled, the blocking of the earth's infrared radiation would rise from 150 to 154 watts per square meter, an increase of roughly 3 percent. This means that the increasing level of carbon dioxide in the atmosphere is not matched by a corresponding increase in the greenhouse effect.

Over the last 100 years, for example, the level of carbon dioxide has increased by 25 percent, while the greenhouse effect has increased by around 1 percent. (This 1-percent figure assumes that other things have stayed equal in the meantime, but in the real world "other things" are usually not so obliging, so the actual behavior of the greenhouse effect during this time is unknown. Nonetheless, its variation has been much closer to 1 percent than to 25 percent.)

Ordinary water vapor is actually the main contributor to the greenhouse effect. The balance between the natural processes of evaporation, which pumps water vapor into the atmosphere, and condensation into clouds, which squeezes it out, sets the level of water vapor in the atmosphere.

This means that the greenhouse effect and global warming are an integral part of the biosphere. They have been around at least since the formation of the first oceans and must therefore have preceded mankind's appearance by a few billion years. Indeed, if there were no global warming, if the earth's atmosphere were perfectly transparent at infrared wavelengths, the planet's average surface temperature would be a brisk zero degrees Fahrenheit, instead of the pleasant 59 degrees that we enjoy. Global warming has been an essential ingredient in the evolution of life on the earth.

Yet the illusion that carbon dioxide is the dominant greenhouse gas is extremely widespread. In an impromptu, totally nonscientific survey, I asked 10 of my fellow astronomers, "What is the major greenhouse gas?" Six said carbon dioxide. One added, "But isn't water vapor in there?" Two said water vapor. And one said, "Don't know."

Evidently a surprisingly large number of astronomers think that carbon dioxide is the major greenhouse gas, despite the fact that astronomers need to know how the earth's atmosphere stamps its spectral imprint on the radiation from heavenly bodies and what gases are responsible. In scientific disciplines that do not deal with the earth's atmosphere on a professional basis, the illusion that carbon dioxide is the major greenhouse gas is probably even more prevalent. Among the general public it must be well-nigh universal.

"Among the greenhouse gases, carbon dioxide gets the lion's share of attention. . . . But it is actually a minor player."

Opposing Mechanisms: Warming Versus Cooling

But even if carbon dioxide is a minor greenhouse gas, its level is increasing. Doesn't this mean global warming is increasing? Yes. But the real question is at what level, and is it significant compared to the changes in global warming that take place independent of mankind's activities? Will it cause an 8-degree increase during the next century, as predicted in the most alarming scenarios, or will there be a much more gradual, and mostly beneficial, increase of 1 degree or so?

We cannot answer this question by means of mere calculation, because our theoretical understanding of the biosphere is too incomplete. The *immediate* result of increased carbon dioxide is, indeed, an increase of global warming. The slightly higher average temperature leads to increased evaporation from the oceans, which leads to a further increase in global warming because water vapor is also a greenhouse gas. But this is far from the end of the story.

More water vapor in the atmosphere leads to increased cloudiness over the earth as a whole. This means more sunshine is reflected straight back into space and so never reaches the earth's surface. This, in turn, means less heating of the earth's surface and hence lower temperatures.

We don't know which one of these opposing mechanisms wins, so we don't know if the increase in the greenhouse effect is amplified or dampened by the time it feeds through to global warming. The availability of faster computers promises that during the next decade we will be able to get a better grip on these factors and many other, more complicated ones that are currently neglected.

In the meantime, the observational evidence that global warming is actually increasing is very shaky. Some researchers claim to see an increase in global warming of about 0.8 degree since 1860. Although average temperatures since then have increased by this much, it's doubtful that the rise reflects carbon dioxide-induced global warming.

The large year-to-year fluctuations of average temperature—which are in the neighborhood of 1 degree—mean that the behavior of average global temperature is somewhat like that of a stock market index. Spotting a "real" temperature trend is like deciding if one is in a bull or a bear market. It's not impossible, but no bell rings when one trend ends and a new one begins.

Natural Causes and Urban Heat

Moreover, natural causes, rather than the increase in carbon dioxide, are a more likely explanation of the temperature increase. Most of the rise took place prior to 1940, *before* the main increase in the carbon dioxide level.

Some climatologists interpret this pre-1940 temperature increase as an aftereffect of the so-called Little Ice Age, a period of unusual worldwide cold that prevailed from 1600 to 1850. If we look at the 50 years or so from 1940 to the present, which have seen the major part of the increase in carbon dioxide concentration, the increase in average global temperatures has been only 0.2 degree. This small increment is within the noise of natural variation. Although the level of carbon dioxide in the atmosphere is increasing, it does not seem to be affecting global temperatures much.

In examining the historical record, we also have to consider the urban heat island effect. The buildings and pavement of a city give it a microclimate slightly warmer than that of the surrounding countryside. As cities have grown, their heat islands have grown with them, so their weather stations, which tend to be in downtown locations, have been more and more prejudiced in favor of higher temperatures.

> ***"Our theoretical understanding of the biosphere is too incomplete."***

Phoenix is a dramatic example. Between 1960 and 1990, as its population grew from 650,000 to 2.1 million, its mean annual temperature heated up by 5 degrees, almost in lockstep with the population increase.

Climatologists who have tried to quantify the urban heat island's influence on the global temperature record estimate that it accounts for somewhere in the neighborhood of 0.2 degree of the 0.8-degree increase seen since 1860. And when Kirby Hanson, Thomas Karl, and George Maul of the National Oceanic and Atmospheric Administration conducted a study of the U.S. temperature record that took into account the urban heat island effect, they found no long-term warming. They confirmed the temperature rise prior to 1940 but found that temperatures have fallen since then.

> ***"The buildings and pavement of a city give it a microclimate slightly warmer than that of the surrounding countryside."***

Another consideration is that most of the earth's surface is covered with water. Temperature data over the oceans is extremely sparse, so the record of the earth's historic average temperature over both land and water is much vaguer than the land record. The recent advent of satellites with the capability to measure global temperatures accurately over both land and sea may solve the problem, but so far their time base is limited to a few years. However, global measurements by satellite of atmospheric (as distinct from surface) temperatures over the last decade show no sign of increasing temperatures.

Crystal Ball Calculations

Far more than historical evidence, the hullabaloo about global warming is based on predictions by computer models. The global climate modeler gives his computer some basic facts, plus a program that recognizes the relevant physical processes and principles insofar as we know them and insofar as they can be calculated. With luck, the computer arrives at a climate not unlike that of the earth. Then the model gets a retroactive "tuning," so that its average global temperature is right.

A calculation of the greenhouse effect and associated global warming is one step in the procedure. Assuming a doubling of atmospheric carbon dioxide, these models predict an increase in global warming of somewhere between 3 degrees and 8 degrees during the next century.

When they are judged by their verifiable accomplishments, however, these computer models are not very impressive. They predict that the temperatures at the poles are lower than those at the equator, and they predict that it's hotter in summer than in winter. But they are weak on specifics. One model predicts an annual rainfall in the central Sahara that is the same as Ireland's.

Clearly, these global climate models are still in a primitive stage of development. They neglect many important factors—both known, such as the poleward transport of heat from the equatorial regions by ocean currents and the atmosphere, and unknown. When it comes to telling us things we don't know already, such as trends in global warming during the next century, they are not far removed from the crystal-ball school of climatology.

Environmental Alarmists

Alarmists such as the vice president are impatient with people who point this out. Gore writes: "If, when the remaining unknowns about the environmental challenge enter the public debate, they are presented as signs that the crisis may not be real after all, it undermines the effort to build a solid base of support for the difficult actions we must soon take. . . . The insistence on complete certainty about the full details of global warming—the most serious threat that we have ever faced—is actually an effort to avoid facing the awful, uncomfortable truth: that we must act boldly, decisively, comprehensively, and quickly, even before we know every last detail of the crisis."

But one of the "details" we still don't know is whether we are in fact facing a crisis requiring drastic action. The burden of proof is on the alarmists. They have failed to meet it.

Urban Smog Levels Are Improving

by K.H. Jones

About the author: *K.H. Jones is president of Zephyr Consulting Co., an environmental consulting firm in Seattle, Washington.*

When the Environmental Protection Agency released its 17th annual report, *National Air Quality and Emissions Trends Report, 1989*, on March 5, 1991, EPA administrator William K. Reilly announced both good and bad news. The good news was that, from 1982 through 1989, atmospheric smog levels fell by 14 percent. The bad news, however, was "the magnitude of the air pollution problem still remaining," given that "sixty-seven million people are living in counties exceeding the smog standard."

Misrepresenting and Manipulating Data

In fact, the March 1991 annual report was but one episode in the EPA's continuing saga of factual misrepresentation and statistical manipulation of the data on urban smog. By refusing to fully acknowledge and appropriately correct for the role of weather in urban ozone trends; by not calling attention to positive preliminary data for 1989, 1990, and 1991 in a timely manner, even though it had [called attention] to negative preliminary data in 1988; by misrepresenting the meaning of ozone "nonattainment"; and by failing to differentiate between distinctly California ozone problems and those of the other 49 states, the EPA has purposefully and cynically misled the nation about the true extent of urban smog. The price of the EPA's misrepresentation will be paid by the American people, who will unnecessarily spend billions of dollars and possibly sacrifice tens of thousands of jobs to solve a problem that exists only in the minds of EPA bureaucrats and environmental advocacy groups.

Urban smog, which is known as ozone pollution, is produced by a complex series of chemical reactions involving automotive and industrial emissions of

From K.H. Jones, "The Truth About Ozone and Urban Smog," Cato Institute *Policy Analysis*, February 19, 1992. Reprinted with permission.

volatile organic compounds (VOCs, mainly hydrocarbons), nitrogen oxides (NO_x) from the same sources, and sunlight. As temperatures increase during the day, solar energy enhances those chemical reactions and increases the amount of ozone produced. Correspondingly, as temperatures decrease, the chemical reactions are slowed and smog is seldom formed. Ozone formation is thus a daytime phenomenon, which occurs during the late spring and summer months in most of the United States.

The health effects of ozone pollution have been studied extensively over the past 20 years. Epidemiological studies have failed to demonstrate any chronic health effects. In 1986 the EPA concluded that

> reported effects on the incidence of acute respiratory illness and on physician, emergency room, and hospital visits are not clearly related with acute exposure to ambient ozone or oxidants and, therefore, are not useful for deriving health effects criteria for standard-setting purposes. Likewise, no convincing association has been demonstrated between daily mortality and daily oxidant concentrations; rather, the effect correlates most closely with elevated temperatures.

The Clean Air Act of 1970 required the EPA to establish a National Ambient Air Quality Standard (NAAQS) for ozone concentration. The standard was originally set at 0.08 parts per million (ppm) but was subsequently revised to 0.12 ppm in 1979. The clinical studies on which the standard is based demonstrated reversible effects above 0.15 ppm. Hence the 0.12-ppm standard incorporates the statutory requirement of an ample margin of safety.

Ozone Monitoring

State and local agencies, with heavy EPA financial and technical support, maintain an extensive ambient ozone monitoring network. There are more than 800 ozone monitors located in 467 counties where ozone has been thought to exceed the 0.12-ppm standard. If even one monitoring device registers an ozone concentration over 0.124 ppm for one hour or more, the entire region is officially found to exceed the NAAQS that day. If any monitor shows four exceedances during any consecutive three-year period, the region is declared an ozone nonattainment region.

> ***"The EPA has purposefully and cynically misled the nation about the true extent of urban smog."***

Nonattainment regions are classified by the EPA as marginal, moderate, serious, severe, and extreme, depending on the concentration of ozone found in the fourth highest reading at any monitoring station over the most recent three-year time frame. The boundaries of nonattainment regions vary depending on the air quality monitoring data.

Thus, the fact that a person lives in an ozone nonattainment region is not an indication of his actual exposure to unhealthy levels of ozone. Almost all exposures of persons living in nonattainment areas outside California are only two to five days more than the permissible one-day-per-year exposure. The maximum exposure concentration on the worst day, with rare exceptions, is within the current legally mandated margin of safety, below 0.15 ppm ozone. In terms of total annual exposure, there is no significant difference between one day and a few days of exposure above the standard. Finally, one monitor does not generally reflect actual ozone exposures over an entire metropolitan area, yet one monitor may detect a one-hour exceedance that causes the whole region to be declared in nonattainment.

"One monitor does not generally reflect actual ozone exposures over an entire metropolitan area."

Finally, it is worth noting how misleading Reilly was being when he said that 67 million people were living in areas that exceeded the ozone standard. Although he was undoubtedly attempting to underscore what he perceives to be the widespread, serious nature of the problem, he failed to mention that 85 percent of all ozone exposures above 0.12 ppm in the nation occur in California and that 82 percent of all exposures nationwide occur in the Los Angeles Basin.

The Great 1988 EPA Smog Scare

Although the EPA has long been relatively silent about the true implications of ozone nonattainment—preferring to dramatize each year's nonattainment readings without providing a proper context—the agency entered an entirely new realm of misinformation in August 1988. For the first time since it began collecting nationwide air quality data from state and local agencies, the EPA publicly released preliminary urban ozone data before the official end of the ozone-monitoring period and some 10 months before the end of the normal data collection cycle. In a memo to Administrator Lee Thomas dated August 18, 1988, EPA's acting assistant administrator for Air and Radiation, Don Clay, noted that the national air monitoring station coordinators had updated most of the summer's ozone data through August 12 and cautioned, "In order to expedite the reporting of the ozone data, it must be clearly understood that the data has not gone through the normal validation process." Neither Clay's memo, nor "A Preliminary Comparison of 1988 Ozone Concentrations to 1983 and 1987 Concentrations, August Update," dated August 15, 1988, and submitted by William G. Laxton of the Technical Support Division to the administrator in preparation for his appearance on the MacNeil/Lehrer Newshour, nor Laxton's cover memo makes any mention of the unusually hot weather that was producing the ozone exceedances.

It would seem that the unprecedented early release of data was politically motivated. The 1988 elections were looming, and then-candidate George Bush was promising that an amended Clean Air Act would be one of the centerpieces of his domestic agenda. Given the unusually hot summer of 1988, ozone readings were reaching record highs in U.S. cities, and alarming smog reports from the EPA would pressure candidate Bush into making campaign promises he would be bound by in the first year of his administration. And just to make sure that the unusually elevated levels of urban smog were a central part of the presidential debate, the Natural Resources Defense Council (NRDC) issued parallel scare reports to back up the EPA's position. Clearly, special interests were at work to manipulate the release of the 1988 data to create a crisis atmosphere that would demand political action.

Weather's Effect on Smog

However, the significance of the EPA's politicization of its preliminary 1988 findings pales in comparison with the agency's failure to explain the true causes of the 1988 ozone readings. Although the EPA pointed out repeatedly that a record number of people were living in nonattainment regions in 1988, a marked deterioration in air quality compared with trends earlier in the decade, the agency failed initially to mention atypical weather as a possible cause of the extremely high levels of smog in the midwestern and eastern United States. Instead, the EPA argued that America was simply polluting too much, that current VOC strategies were not adequate, and that stringent new federal regulations were needed to prevent air quality from getting progressively worse.

> ***"It would seem that the unprecedented early release of [ozone] data was politically motivated."***

The fact is, however, that weather—not VOC or NO_x emissions—is the most important determinant of the number of days a year that are conducive to ozone formation. Judging by the number of days on which the temperature rose above 90 degrees Fahrenheit, the summer of 1988 was among the hottest two or three summers in the past 25 years.

Before the 1988 preliminary ozone data release, the consensus thinking in Washington had been that there would be some relaxation of the most stringent regulations in the upcoming amendments to the Clean Air Act relative to ozone nonattainment strategies (e.g., making state automobile inspection and maintenance programs discretionary rather than mandatory). Few people expected Congress to try to "fix" an ozone nonattainment problem that was well on the way to being solved. Thanks to the EPA's and the NRDC's intentional misinterpretation of the 1988 data, however, we now have a more draconian and expensive urban ozone regulatory program than was imaginable in 1988.

In response to scattered criticism of the EPA's alarmist 1988 ozone findings, the agency argued that, although the summer of 1988 was indeed unusually hot, hotter summers were a pronounced trend in the 1980s and the "greenhouse effect" might increase the likelihood of even hotter summers in the future. Thus, according to the EPA, using the 1988 data to establish nonattainment regions was perfectly legitimate and in fact prudent if America was to ensure clean air in the future.

To test the EPA's greenhouse assertion, I compiled an annual 90°F temperature profile for more than 90 ozone nonattainment areas for 1967 through 1989. That temperature yardstick is a good indicator of how many days per year weather conditions (i.e., high temperatures, low wind speeds, and temperature inversions) are conducive to smog formation.

The median number of days above 90°F [in seven cities surveyed] during the first half versus the second half of the 25-year period demonstrates that there is no trend of increasing temperatures for the cities that suffered from record temperatures during the summer of 1988. Three of the six areas influenced by the drought (all but Dallas/Fort Worth) show an increase, and three show a decrease in the median for the first versus the second time period. Other cities in those regions show the same pattern. . . .

The EPA Drags Its Heels

In August 1988 the EPA proved that it could act rapidly to report ominous air pollution data. However, in subsequent years, it made no effort to call the public's attention to encouraging preliminary data.

The preliminary data for 1989 indicated a dramatic reduction in urban smog levels from 1988 and a return to expected ozone levels, but the EPA did not call public attention to those data. It waited until March 1991 to issue the annual report. When I asked the EPA, in the fall of 1989, why the good news about the preliminary data was not being publicized, the answer was, "No one has asked us for the data."

> ***"Weather . . . is the most important determinant of the number of days a year that are conducive to ozone formation."***

The 1991 ozone readings were as favorable as the 1990 readings. Their incorporation into a new three-year ozone nonattainment data base, from which the aberrant 1988 readings would be removed, would radically alter the ozone nonattainment status of dozens of cities. Although EPA staff members concede that they have indeed gathered the preliminary 1991 data and confirm that there *was* a distinct downward trend in national nonattainment status, the agency argues that it cannot release a preliminary report because "the 1991 data have yet to be quality assured." Of course, the agency did not let quality assurance concerns get in the way of its rushed 1988 preliminary report. Given the tens of billions of dollars that would be saved by immediately adopting the 1989-91 data base, it is outrageous that the EPA refuses to spend the few thousand dollars necessary to provide quality assurance for the already compiled 1991 numbers.

During the debate on the 1990 amendments to the Clean Air Act, the EPA steadfastly supported the use of the 1988 data in classifying nonattainment areas. The establishment in the amendments of a closure date of November 1990 for data review for baseline nonattainment designation and severity classification ensured the worst possible portrayal of the nation's ozone problem. The EPA could and should have proposed to Congress a one- or two-year wait-and-see strategy to ensure scientific honesty.

Now the act requires states to go through several arduous steps to get redesignated or reclassified, or both, on the basis of the new data. It does not appear that the EPA is encouraging the states to do so.

1991 Ozone Findings

Even if the nonattainment data are not adjusted to account for temperature fluctuations, recalculation of regional nonattainment status using the latest three-year cycle (1989-91) reveals a dramatic improvement in urban air quality. . . .

Examination of EPA data reveals that there has been more than a 60 percent nationwide reduction in median ozone exceedances since 1988. That dramatic improvement in ozone air quality was obviously not due to increased regulatory activity alone, although the EPA may now wish to make such a claim. More important, there are only 28 nonattainment areas outside California, not 89 as reported by the EPA in November 1991. That dramatic reduction in urban ozone is even more remarkable when one considers that the 1989-91 data have not been adjusted for temperature, and it can be readily demonstrated that the most recent data have a slight hotter-than-normal bias.

Table 1 is a subset of the EPA data. It shows the annual number of nonattainment areas and the total number of exceedances nationwide. The data for California are separated from those for the rest of the nation.

Several major conclusions can be drawn from the data.

- Non-California data for 1988 were clearly aberrant.
- There is a clear downward trend in ozone pollution (in terms of both total exceedances and number of nonattainment areas) when the 1988 data are excluded from the non-California trend data base.
- California stands alone as an ozone regulatory problem, principally because of the Los Angeles Basin.

The last conclusion makes a mockery of the EPA's implication that Los Angeles's regulatory strategies should be applied to other regions of the nation. It must also be remembered that California has an ozone standard of 0.10 ppm,

Table 1

Annual Nonattainment Rates and Total Nationwide Exceedance Levels, 1985-91

Case	Year						
	85	86	87	88	89	90	91
Number of Areas with 2 or More Exceedances							
California	9	8	8	9	9	5	6
Other U.S.	33	36	43	85	20	20	29
Total U.S.	42	44	51	94	29	25	35
Total Number of Exceedances							
California	257	272	270	308	233	167	90[a]
Other U.S.	188	166	275	607	101	124	155
Total U.S.	445	438	545	915	334	291	245
Percent Calif.	58	62	30	34	70	57	37

[a]Data on Los Angeles are incomplete.

which inevitably makes comparisons between California and the rest of the nation problematic. . . .

Summary: The EPA Must Come Clean

Reilly's rhetoric aside, America has made great strides in smog abatement over the past decade. Temperature-adjusted data indicate that ozone pollution outside California has been reduced by 74 percent since 1985. Today only three urban areas outside California have serious or severe ozone nonattainment problems. Another 25 areas, which suffer only marginal to moderate smog problems, show every sign of achieving attainment within two to five years without the additional onerous regulatory controls spelled out in the 1990 amendments to the Clean Air Act. Any major ozone-smog problem in America is confined to the state of California, particularly the Los Angeles Basin. It is ridiculous to treat all of America as if it faced the problems California does and to impose on the entire nation massive economic costs that are ultimately unnecessary and counterproductive.

In pursuit of greater budgets, increased regulatory authority, and the political benefits of front-page coverage, the EPA has perpetrated a fraud on the American people. The agency's refusal to acknowledge that the 1988 data on ozone were an aberration and its failure to publicize preliminary 1991 data in a timely manner could cost the economy $26 billion a year. The result can only be continued economic stagnation, higher unemployment, and reduced international competitiveness.

Chapter 2

Are Corporations Polluting the Environment?

Industry and Pollution: An Overview

by S. Prakash Sethi

About the author: *S. Prakash Sethi is a senior editor of the quarterly journal* Business and Society Review. *He is also an associate director of the Center for Management at Baruch College of the City University of New York.*

In October 1990, the U.S. Congress finally passed, and President [George] Bush signed, a new Clean Air Act. It sets forth stringent conditions on a wide variety of emission sources, making it, from all accounts, the most expensive environmental law ever enacted by Congress. It is estimated that the new Act will cost businesses, consumers, and taxpayers between $20-$50 billion annually. The funds will be spent on new fuels, pollution-control equipment, and new industrial processes, with the objective of making the air we breathe gradually cleaner over the next fifteen years.

Environmental Rifts and Catastrophes

The Act was thirteen years in the making. This period was marked by extreme acrimony between pro- and anti-environmental groups, between moderates and radicals, between compromisers and purists, and between those who seek short-term benefits and those who would impose greater sacrifices on the current generation in order to secure a greener, more pristine environment for future generations. This period also witnessed some major environmental catastrophes that caused deep and often irreparable harm to human life and habitat, deterioration of the fragile ecosystem, and costing untold millions to the communities and companies involved. In the process, Union Carbide's Bhopal, Exxon's *Valdez*, and Love Canal and Times Beach have become unforgettable and unforgivable members of the Environmental Hall of Horrors. In addition to the Clean Air Act, this period also saw the passage of a number of other environmental laws, including the Comprehensive Environmental Response, Com-

From S. Prakash Sethi, "Corporations and the Environment: Greening or Preening?" *Business and Society Review*, Fall 1990. Reprinted with permission.

pensation, and Liability Act (CERCLA). That law, commonly known as Superfund, has apparently so far generated more legal action and lawyers' fees than cleaned up toxic waste sites.

One of the most surprising aspects of the Clean Air Act's legislative history was that it brought together competing interests that eventually fashioned an acceptable, workable compromise. There was not, of course, a complete meeting of minds, but a certain *modus vivendi* was achieved; it was a most positive demonstration of the democratic system at work. Business lobbyists did not bitterly complain about the excessive costs of compliance, lack of available technology, potential loss of international competitiveness, and environmentalists running amok. Nor were environmental activists accusing business of backroom politics, obstructionist tactics, political blackmail through campaign contributions, and putting profits before the environment and public health.

"It would seem that the greening of America has finally arrived, albeit a bit late."

It would seem that the greening of America has finally arrived, albeit a bit late. The issue is no longer whether the environment needs protection, but rather what are the most efficient ways of spending the money and how those costs should be equitably apportioned among groups.

Part of the Solution

Efforts in the 1990s to achieve environmental goals will present business with great challenges—and also afford it great opportunities. Business will not be viewed solely as part of the problem but instead as an important part of the solution. For without the cooperation of industry, the world would not grow cleaner or safer. While industry can wreak havoc on the environment, it does so partly because society created the structural conditions and product demands that made such environmental degradation financially attractive and socially feasible. Therefore, given the right legal and political signals, altered consumer demands, and appropriate market incentives, industry can also create new technologies, manufacturing processes, and end-use products that are environment friendly.

The evolution of the new environmentally sensitive era has been a long time coming. The environmental battles of the 1970s and, most notably, the 1980s, suggest a confluence of major forces that brings us where we are today:

• *Greening of politics.* Public opinion indicates increasing support for environmental issues. Government has responded with more intervention in the marketplace to require companies—and consumers—to meet rigid environmental standards throughout the chain of production and consumption. Grassroots opposition to any industrial activity that may pose environmental hazards—the "not in my backyard" mentality—has made it increasingly difficult for companies to establish new plants or waste-disposal sites. At the national level, political leaders and regulatory authorities are increasingly willing to take the initiative and set toxic emission standards, tax raw materials, proscribe uses of potentially harmful materials, assign liability for pollution law violations, and dictate new technologies—not only to correct past mistakes but, more important, to prevent future environmental hazards.

• *Greening of the pocketbook.* Consumers have demonstrated an increasing willingness to prefer products and services that are not harmful to the environment. Although the sustaining power of this movement is uncertain, there is no question about its current assertiveness. Consumer groups have resorted to product and company boycotts, waste-reduction campaigns, public education, and investing in environment-friendly companies. . . .

The Need for Protection

• *Greening of industry.* Industry itself has come to recognize the need for environmental protection because of the triple threat of increased costs, public hostility (and therefore regulation), and potential liability from environment-related accidents affecting consumers, workers, and communities. This heightened recognition is best illustrated by an observation of Bjorn Stigson, head of AB Flakt, a Swedish engineering firm: "We treat nature like we treated workers a hundred years ago. We included then no cost for the health and social security of our workers in our calculations, and today we include no cost for the health and security of nature." Companies such as Du Pont, Monsanto, Union Carbide, and Occidental Petroleum have developed elaborate and detailed standards and internal audit programs to monitor compliance with both state-mandated and company health, safety, and environment regulations and standards. Monsanto, for instance, pledged to cut its toxic waste emissions 90 percent by 1992. Such environmental actions can also be profitable: 3M, for example, claims to have saved over $480 million worldwide through its "Pollution Prevention Pays" program.

The path to a better and healthier environment, however, is not likely to be smooth or trouble free. According to the *Economist*, there are no more than 100-200 companies worldwide that have made environmental performance one of their top concerns. A great many others merely pay lip service to the environment or have jumped on the bandwagon by taking out glossy advertisements touting their alleged environmental performance. All of this is inevitable. Thus, government, public interest groups, and consumers will have to maintain their vigilance and work in cooperation with forward-looking industries and enterprises to ensure that environmental gains are substantive and not media glitz. This is the challenge of the 1990s.

> ***"Industry can . . . create new technologies, manufacturing processes, and end-use products that are environment friendly."***

Large Corporations Are Serious Polluters

by Colin Crawford

About the author: *Colin Crawford is a senior researcher for the Council on Economic Priorities (CEP), a public interest research organization in New York City that evaluates the policies of U.S. corporations and issues affecting national security.*

In 1970 the *Washington Post* called CEP the "wheels on which the vehicle of active protest can move." CEP's Campaign for Cleaner Corporations (C3) shows that the *Post*'s assessment remains as true today as it was twenty-two years ago. Made possible by a generous donation from Alida Rockefeller, C3 is an innovative effort to identify some of the nation's worst corporate environmental performers by comparing them to other companies in the same industry. The Campaign's research will be used by a wide range of environmental and other activist organizations. Groups that plan to use our work range from grassroots activists with chapters nationwide, such as Citizen Action, to highly-respected research enterprises like Lester Brown's Worldwatch Institute and national membership organizations like the Sierra Club.

Ranking Corporate Polluters

C3 will also make its research available to shareholder groups, such as the Social Investment Forum, to help shape investment and shareholder action decisions. Ariane Van Buren, Director of ICCR's Energy and Environment Program and a veteran of several socially minded shareholder resolution campaigns, said "CEP's efforts are useful to us because they provide fuller objective reports on companies than are otherwise available." She added that the "campaign draws attention both to the availability of that information and to the efforts that shareholders can make to raise the level of accountability of U.S. corporations

From Colin Crawford, "Campaign for Cleaner Corporations," *Council on Economic Priorities Research Report*, December 1992. Reprinted with permission.

for their effect on the environment."

Companies were selected for the campaign by a blue-ribbon panel of judges including prominent scientists, academics, environmentalists and investors.

C3's goals are straightforward. The Campaign aims to increase public awareness of the comparative environmental performance of large U.S. companies in a variety of environmentally hazardous industries. The Campaign will do this by publicizing our research on corporate environmental performance. We hope that ensuing public pressure on C3 companies will result in more responsible corporate environmental practices.

"C3 is an innovative effort to identify some of the nation's worst corporate environmental performers."

Here are some disturbing details from the environmental records of the eight C3 companies, listed in alphabetical order:

Cargill. The world's largest grain trader and tenth largest seed company, and the nation's second or third largest meat packer and third largest flour miller is also one of the most secretive corporations in the U.S. Nonetheless, CEP researchers documented Cargill's unenviable environmental record. For example, Cargill precipitated a world-class environmental disaster when in 1988 it spilled 40,000 gallons of phosphoric solution into Florida's Alafia River from its Gardinier phosphate facility in Gibsonton, Florida. After stopping this marine life-killing activity, the Cargill facility spilled another 300,000 gallons of phosphoric acid onto the land between October 1989 and September 1990.

Cargill was cited in 1988 in the Toxic Release Inventory (TRI) of the U.S. Environmental Protection Agency (EPA) as the agricultural products company that released the largest amount of toxic substances. In 1989 (the last year for which data is available) it was the second largest. It has the worst air pollution compliance record of any major company in its industry. Cargill also has one of the most egregious occupational safety records of any major agricultural products company. Since 1987, Cargill has been cited by the U.S. Occupational Safety and Health Administration (OSHA) for over 2,000 violations. Many of them involved "knowing and willful" exposures of employees to workplace hazards.

A Toxic Polluter

Du Pont. In recent years, Du Pont CEO Edgar S. Woolard, Jr. has waxed eloquent about the need for corporations to clean up their acts. However, Mr. Woolard has a lot yet to do at his own company. EPA data indicates that in 1989 the company released toxics at the rate of more than a million pounds a day. The same year, Du Pont led all other U.S. companies in the controversial practice of deep-well injection of toxic wastes.

Despite efforts to develop chlorofluorocarbon (CFC) substitutes, Du Pont remains the world's largest producer of the ozone-depleting substances. Moreover, Du Pont's problems are not confined to the U.S. The company has been cited for its policy of exporting hazardous wastes, and it does not require its numerous overseas operations publicly to report toxic releases.

General Electric. General Electric is responsible for the aggregate release of more toxic chemicals than any other company in the electrical industry. Some parts of the country have been especially hard hit by GE's negligent environmental practices. Over a 30-year period, GE discharged about 500,000 pounds of polychlorinated biphenyls (PCBs) into the Hudson River.

Many of GE's environmental problems involve faulty products. Several utilities have filed suit against GE, alleging fraud and negligence for providing nuclear-containment vessels that needed excessive service or repairs. In 1989, GE was criticized for allowing CFC-laden coolant from 300,000 defective refrigerators to escape into the atmosphere. Moreover, GE has been cited for more OSHA violations than any company in its industry for the 1987 to 1992 period.

Despite its environmental problems, GE markets itself as an environmentally friendly company. For example, GE's exaggerations about its "Energy Choice" lightbulbs have earned it the dubious honor of awards for misleading, unfair or irresponsible advertising.

"Cargill was cited . . . as the agricultural products company that released the largest amount of toxic substances."

General Motors. The nation's largest manufacturer is also a major polluter. In 1988 and 1989, GM released nearly three times as much toxic material as Ford, its principal competitor. GM is a potentially responsible party at about 200 Superfund sites. Some of these are among the worst in the nation. One such site is the polychlorinated biphenyl (PCB)-polluting GM facility in Massena, NY. The Massena site has contaminated an area that extends into Canada and includes portions of a Mohawk reservation, the St. Lawrence and other rivers. The Massena site is expected to cost over $100 million to clean up.

Another GM site is alleged to have caused ground contamination by emitting large quantities of toxic PCBs near Michigan's Saginaw River. State regulators are seeking over $32 million in penalties.

In California, GM has been named the state's largest producer of ozone-depleting chemicals.

The automaker paid the largest total penalties for federal and state OSHA violations during the 1982-1990 period.

Devastating the Forests

Georgia Pacific. Georgia Pacific (GP) is the nation's largest importer of timber from tropical rainforests. In this country, GP has been accused frequently of overharvesting timber to pay off a substantial corporate debt. Its overharvesting in places like the Tacoma, WA, area caused serious erosion and may have destroyed ecosystems. Among the tree stands that it has overharvested are old-growth forests containing virgin redwood. California environmental groups have obtained court orders forcing the company to cut back on its harvesting plans.

Of the 20 forest products companies studied by CEP, GP had the worst record of air compliance and has been named as one of the 25 U.S. companies that emit the most carcinogenic substances.

Numerous lawsuits allege billions in damages by GP as a result of dioxin discharge into Mississippi's Leaf, Pascagoula, and Esctawpa rivers. Additional lawsuits are pending in Mississippi and Texas. GP claims that it is reducing its emissions of the highly dangerous substance.

Maxxam. With $2.3 billion in sales in 1991, the conglomerate (built in the 1980s by corporate raider Charles Hurwitz out of metals and timber acquisitions) is smaller than other companies on the C3 list, but the extent of its environmental irresponsibility is as great. Pacific Lumber had been called a model of an environmentally aware logging company before MAXXAM's hostile takeover. Now it is infamous in California and the Pacific Northwest, where it continues to cut virgin old-growth redwood forests. In 1990, 75% of MAXXAM's lumber production was redwood, of which nearly 40% was virgin old-growth timber. By 1991, MAXXAM was logging at nearly double its 1985 (pre-buyout) rate.

MAXXAM shows little desire to improve its environmentally dangerous practices and lacks a formal environmental policy. Although its Kaiser Aluminum subsidiary releases a large volume of toxics (adjusting for sales, it had the fourth worst TRI figures in the metals industry in 1989), MAXXAM is one of the few companies with such operations that has not joined EPA's Industrial Toxics Program, a voluntary initiative to reduce by 50% the industrial release of 17 toxic chemicals by 1995.

Radioactive Hazards

Rockwell. EPA has indicated that Rockwell's Rocky Flats, CO, nuclear weapons manufacturing facility may be the nation's most polluted site. EPA documented 166 separate hazardous waste dumps at Rocky Flats. Studies have linked plutonium exposure to cancer among workers and residents living near Rocky Flats.

"Rockwell's Rocky Flats, CO, nuclear weapons manufacturing facility may be the nation's most polluted site."

Radioactive contamination is also a major problem at other Rockwell sites. At California and Washington facilities, poor technology design has been cited as the reason for worker exposure to plutonium and other radioactive materials.

For several years, the aerospace and defense manufacturer was named the second-greatest emitter (after GM) of ozone-producing chemicals in California—3.5% of the state's total in 1990 (the last year for which figures are available). In California's San Fernando valley, 20 years of radioactive material releases may have caused high rates of bladder cancer, according to *Corporate Crime Reporter.*

> ***"Du Pont remains the world's largest producer of the ozone-depleting substances."***

USX. USX—which includes U.S. Steel and Marathon Oil—has long had an especially bad environmental record in air and water pollution and toxics control. For example, USX was forced to close down most of its Fairless, PA, steel works in 1991 due to repeated air pollution control violations. EPA also sued USX in 1991 for Clean Air Act violations in Clairton, PA. In 1989, USX paid $125,000 for violations at that facility.

The company's water pollution record is particularly lamentable in Gary, IN. For dumping toxic wastewater at that facility USX has paid over $34 million in EPA fines. Because of illegal dumping at its Gary facility, EPA suggested barring USX from government contracts—its most severe penalty.

USX also has one of the nation's worst occupational safety records. From 1977 through 1990, USX paid the eighth largest total of OSHA penalties among the nation's 50 largest manufacturing companies. OSHA reported that between 1972 and 1989, 17 workers were killed on the job at USX facilities in Pennsylvania alone.

Foreign Companies and the Oil Industry

The judges expressed two additional concerns. First, they observed that little information is available on corporate environmental practices abroad. For example, judges cited the poor foreign environmental record of the Swiss-based chemical manufacturer Ciba-Geigy, which has allegedly sprayed pesticides on human subjects in Asia and Africa (as reported in the respected newsmagazine *India Today*) and routinely exports large quantities of pesticides manufactured in, but banned for use in, the U.S. In Europe, Ciba-Geigy's air emissions of poisonous gases and water contamination by inadvertent releases of pesticides and other chemicals have repeatedly caused local health disasters, such as a 1986 cloud of 1,100 pounds of toxic phenol released over Basel, Switzerland. The limited data available in the U.S about these alleged incidents led the judges to stress the need to develop a uniform, global system for corporate disclosure of environmental practices and records.

Second, the judges cited large U.S.-based oil companies for poor environmental records. Although they were unable to agree as to which oil company had the most egregious environmental record, they expressed the hope that the entire oil industry will improve.

A press initiative will spotlight the eight companies whose practices and operations were deemed to be especially environmentally hazardous.

A story on the Campaign will appear in the January [1993] issue of *Mother Jones* magazine. Editor David Weir said that *Mother Jones* was interested in the Campaign because it "introduces hope." Weir continued: "CEP has performed an important public service by putting together publicly available environmental reports in a responsible, accurate manner." Weir explained that too often the magazine receives piecemeal information that is difficult to evaluate. He likes CEP's reports because they feature comparative industrial data, making it easier to mark corporate progress, and also because they indicate efforts companies are making to improve their environmental performance.

Although C3 identifies companies that have been environmental villains, its purpose is ultimately to work for positive change. CEP hopes that by isolating the worst environmental performer in an industry, C3 will serve as a catalyst for improvement. Even among the eight companies with egregious records, positive changes are beginning to take place.

U.S. Factories in Mexico Cause Toxic Pollution

by Daniel Brook

About the author: *Daniel Brook is a doctoral student in political science at the University of California in Davis.*

The governments of Mexico and the United States initiated negotiations in June 1991 on a Free Trade Agreement [FTA], a bilateral version of the General Agreement on Tariffs and Trade [GATT], that will, in the long run, have profound effects upon the political economy of each nation. The U.S., already engaged in FTA's with Israel (1985) and Canada (1989), concluded negotiations with Mexico during 1992, ironically coinciding with the Columbus Quincentennial. This major step has serious implications for all parties involved—governments, capitalists, and workers alike—in addition to the environment. An FTA between countries of such different socioeconomic levels, i.e. First World and Third World, is unprecedented. For some it will be a miracle, for others a mirage, and still others a menace.

Scope of the Free Trade Agreement

The agreement between the U.S. and Mexico aims to eliminate all restrictions that hamper the free flow of goods, services, and investments across the political fiction of the international border, of which this is the longest separating the First and Third Worlds. Restrictions on intellectual property, however, will be increased, as 92 percent of patents registered in Mexico are already foreign owned. Interestingly, foreign debt, one of Mexico's chief obstacles to development, and oil, tourism, and migrant workers, three of the four highest foreign exchange earners for Mexico, will not be addressed in the FTA. Neither will environmental concerns, workers' rights, health and safety, wage and standard of living discrepancies, migrants and labor mobility, consumer protection, nor will political democracy be a part of the FTA. . . .

From Daniel Brook, "Toxic Trade," *Z Magazine*, September 1992. Reprinted with permission.

The FTA is deregulation par excellence in that it seeks to harmonize standards to their lowest common denominator and lessen restrictions on transnational economic activity. The lowering of standards has already started taking place in North America. For example, "the U.S. . . . criticized Canadian acid rain pollution laws as an unfair trading practice." The U.S. has even attacked Canada's national health care system as an unfair trading practice, arguing that it constitutes a governmental subsidy to business. Meat inspection and asbestos control by the U.S. was considered a trade barrier by Canada. Mexico is trying to persuade the U.S. to repeal its dolphin-free tuna importation policy for the same reason. And, the crucial point, as Ralph Nader and Michael Waldman argue, is that "in recent years the U.S. has changed its laws every time it has been challenged" for unfair trading practices with only one exception. Dispute resolution rules prohibit consideration of "any factors other than strictly commercial ones. Arguments about consumers' health, worker safety or environmental impacts would not be allowed." Conservatives in the U.S. government are accomplishing through treaties and international agreements what they are not able to do through legislation. Under GATT and the FTA, Kathryn Collmer, a food and agriculture policy analyst, explains that the U.S. "would be forced to import fruits and vegetables contaminated with pesticide residues much higher than the accepted U.S. levels, including pesticides such as DDT that have been banned in the U.S."

> ***"DDT [insecticide] residues on imported food from Mexico could . . . increase by as much as 5,000 percent."***

A Threat to Food Safety

In fact, DDT residues on imported food from Mexico could, according to Craig Merrilees, National Co-Director of the Fair Trade Campaign, increase by as much as 5,000 percent on such produce as peaches and bananas, for example. Other pesticides, including aldicarb, aldrin, benomyl, diazinon, endrin, heptachlor, lindane, and permathrin, will be allowed to be residually present on imported fruit by as much as 160-4,000 percent higher than current permissible levels. Collmer additionally states that:

- We would be forced to import meat and dairy products with high levels of hormone residues and disease contamination.
- Restrictions on pesticides and hormones in domestic farm products would be loosened, so that U.S. farmers could compete more effectively against cheap imports.
- Restrictions on the sale of milk containing genetically engineered bovine growth hormone (BGH) would be eliminated.
- Rules that require irradiated foods to be labeled as such would be abolished.

• The Food and Nutrition Labeling Act of 1990 would be repealed.

• The Delaney Clause of the Food, Drug and Cosmetic Act, which prohibits food additives that cause cancer, would be repealed.

• Proposed "Circle of Poison" legislation, which would prohibit U.S. chemical companies from exporting pesticides that have been banned in the U.S., would be killed.

The Maquiladora Industry

As the maquiladora industry [foreign companies operating factories in Mexico], a model for North American free trade, opened the border up to foreign capital investment, so too did the border increasingly open up to environmental degradation. Many of the maquiladoras find it more profitable to pollute and contaminate Mexico than to properly dispose of their waste. The poverty of the people exacerbates the problems of environmental degradation and health hazards. People living near the maquiladoras have been known to store their water in carelessly discarded metal drums that once held toxic chemicals. Most of them are unable to read the English warning labels, some are too desperate to care, others too weak from malnourishment to notice. Environmentalists have estimated that the cost of cleaning up the border region from the pollution and contamination of the air, ground, and water by the maquiladoras would cost much more than the total earnings of the industry, perhaps in excess of $50 billion. This amount does not even include the costs of medical care required by the workers in the plants and the people who live in the area and have been affected by the noxious maquiladoras.

"Many of the maquiladoras find it more profitable to pollute and contaminate Mexico than to properly dispose of their waste."

The maquiladora industry is known to contribute heavily to the destruction of the environment on both sides of the border. In 1987, tacitly acknowledging a very serious problem, Presidents Miguel de la Madrid and Ronald Reagan signed an executive agreement mandating that U.S. owned maquiladoras return their waste products to the U.S. for disposal. Despite this formal agreement at the highest level, the waste has not been returned to the U.S. Indeed, the maquiladoras, a virtual who's who of the Fortune 500, "have poured chemical wastes down the drains, dumped them in irrigation ditches, left them in the desert, burned them in city dumps and turned them over to Mexican recycling plants that are not qualified to handle toxic wastes." In a survey of 772 maquiladoras in the city of Tijuana taken in 1986, it was revealed that only 20 of them had notified the U.S. Environmental Protection Agency (EPA) that they were properly returning their waste to the U.S. The toxic waste that is not returned shows up in the border region in various detrimental ways, poisoning the Earth and its inhabitants. In 1989,

175 drums of PCB's were discovered only two blocks from the Ciudad Juarez-El Paso border. Sadly, this is not an isolated incident.

Toxic Waste in the Border Area

Toxic waste has shown up in significantly high concentrations in many of the waterways that pass through the border area. The Rio Grande/Rio Bravo, though it may be better known, is not the only heavily contaminated waterway. The Tijuana River is the recipient of at least 12 million gallons of toxic chemicals and raw sewage each day. In Mexicali, the New River, which crosses the border and goes through Calexico and the Imperial Valley, contains over 100 types of industrial waste. This has even caused the deaths of fish and birds in Salton Sea, the largest lake in California. The New River is so toxic "that signs warn people to not even go near the river." This river is considered the dirtiest in the U.S., perhaps in all of North America. The White House has even called the New River "one of the most polluted rivers in the world." The river has a thick foam and is full of bacteria, industrial chemicals, and heavy metals. Indeed, "of 16 volatile organic compounds known to be used in the assembly of printed circuit boards, 13 have been detected repeatedly in the water of the New River." The Nogales Wash, another waterway polluted by the maquiladoras, flows from the state of Sonora into Arizona, poisoning low-income people in both countries. Scientists with the Boston-based National Toxic Campaign Fund took samples near the maquiladoras and "found evidence that 75 percent of the sites were discharging toxic chemicals directly into public waterways." One of the perpetrators of these deadly crimes is a maquiladora called Rimir, fully owned and operated by General Motors, one of the three top private sector exporters from Mexico to the U.S. Samples taken near Rimir GM show measurements of the toxic solvent xylene 6,300 to 6,800 times the Mexican and U.S. legal limit. Xylene has been linked to anencephaly, an absence of part or all of the brain in newborn babies. This tragic condition has disproportionately afflicted families on both sides of the Texas-Mexico border. An employee at the Rimir GM plant says "that the company regularly pours untreated solvents right down the drain." Many other big name corporations, including Ford, have had toxic chemicals such as xylene and styrene, among others, detected around their plants. There are ditches around the shantytowns in Matamoros, across from Brownsville, Texas, "that are coated with an iridescent slick of aromatic chemicals, many of which are known or suspected carcinogens." When the coating on these ditches is pierced, "black globules bubble [up] from the bottom, releasing an eye-stinging chemical stench."

> ***"The Tijuana River is the recipient of at least 12 million gallons of toxic chemicals and raw sewage each day."***

Some corporations are relocating their production facilities to Mexico solely to take advantage of the cost effectiveness of discharging their waste products however and wherever they please. A case in point which illustrates this phenomenon is the furniture industry. Up until 1988, Los Angeles was the second largest site for furniture manufacturers in the U.S. This $1.3 billion industry employed 63,000 workers. That year, the South Coast Air Quality Management District mandated the use of spray chambers in all furniture manufacturing plants within its jurisdiction. Shortly thereafter, over 40 firms, constituting the bulk of the industry, announced their plans to relocate in northern Mexico, where the requirements for spray chambers are not enforced. The U.S. General Accounting Office (GAO) reported that 78 percent of the furniture manufacturers cited environmental regulations as a major reason for moving to Mexico. These new maquiladoras still manufacture furniture (at one-tenth the wage levels) and still produce the hydrocarbon fumes from paint solvents, which can be detected blowing back into southern California. If corporations find that it is more profitable to pollute in Mexico, many will do so, for that is the logic of capitalism. A confidential Mexican cabinet-level report concedes that "if it is possible to save money by improperly disposing of dangerous wastes, industry will do it." [*Los Angeles Times* reporter] Juanita Darling states that "the report indicates that high government officials [in Mexico] privately share many of the worries about free trade expressed by ecologists throughout North America."

Dirtying Poorer Nations

Lawrence Summers, chief economist of the World Bank, however, does not share the same environmental worries, even in private. Summers, in a memorandum to his colleagues, asks "just between you and me, shouldn't the World Bank be encouraging more migration of the dirty industries to the LDC's [less developed countries]?" He explains that "a given amount of health-impairing pollution should be done in the country with the lowest cost, which will be the country with the lowest wages. I think the economic logic behind dumping a load of toxic waste in the lowest-wage country is impeccable and we should face up to that." He continues to say that he has "always thought that under-populated countries . . . are vastly under-polluted; their air quality is probably vastly inefficiently low [sic] compared to Los Angeles." World Bank Chief Economist Lawrence Summers concludes by revealingly complaining that "the problem with the arguments against all of these proposals for more pollution in LDC's (intrinsic rights to certain goods, moral reasons, social concerns, lack of adequate markets, etc) could be turned around and used more or less effectively

> ***"Seventy-five percent of the sites were discharging toxic chemicals directly into public waterways."***

against every [World] Bank proposal for liberalization." *The Economist*, which published this memo, incredulously editorializes that although "the language is crass . . . his points are hard to answer." Summers's language, however, is the least crass aspect of this neo-fascist memo.

Researchers believe that exposure to many of the toxic chemicals used and dumped by the maquiladoras "can lead to brain, lung, liver, and kidney damage," in addition to various cancers. *The Journal of the American Medical Association* published a report that in San Elizario, Texas, 35 percent of 8-year-old children had been infected with hepatitis A and by 35 years of age, 85-90 percent of the town's population had been infected. These frightening data are the results of poverty and pollution. In fact, "the three cities farthest east—[on the U.S.-Mexico border, the Texas cities of Laredo, McAllen, and Brownsville]—are not only the poorest in the border region, but are the poorest metropolitan areas" in the U.S. An AMA [American Medical Association] scientific panel has called the border area "a virtual cesspool and breeding ground for infectious disease." It further concluded that a continuation of the present trend of "uncontrolled" pollution caused by deregulated business is severely threatening "the health and future economic vitality on both sides of the border."

Approximately one-third of maquiladora workers complain about serious physical ailments, such as severe back pain, bronchitis and asthma, conjunctivitis and loss of vision. In the electronics maquiladoras, where the assemblers must fuse components together on a tiny circuitry board, workers are required to have perfect vision in order to be hired. It has been shown that after only a few years most of the workers no longer have such good vision, but instead, many have poor eyesight, and indeed, a large number become blind. Many workers have had fingers amputated and others have been subject to chemical burns. The industrial death rate is very high. Psychological problems, including depression, are also common. Sexual harassment, abuse, and rape are rampant. All of these problems have almost invariably been blamed on the victims, as opposed to the exploitative nature and structure of the transnational corporations.

All of the toxic contamination and its ramifications are the daily by-products of the free trade policies of the maquiladora industry along the border which, according to Roberto Sanchez, director of Environmental and Urban Studies at the Colegio de la Frontera Norte, "has essentially been a free trade zone for decades." Sanchez notes that 13 percent of respondents to a recent survey said that "weaker environmental controls in Mexico were a primary factor in relocating to the border area." Therefore, Mexico has become the legal loophole for corporations seeking to evade the hard fought gains of environmentalists, consumer advocates, unions, and social activists. Sanchez reminds us that, despite the announcements of government officials and TNC [transnational corporation] leaders, "the economy is not separate from the environment." Free trade is anything but free and, indeed, it is toxic.

The Plastics Industry Pollutes Third World Countries

by Anne Leonard

About the author: *Anne Leonard campaigns against the waste trade for Greenpeace, a prominent environmental organization that opposes nuclear energy and activities that harm the environment.*

In the 90-degree heat, women stand over huge piles of plastic garbage. It is too hot to wear a protective smock—not that one is available anyway. They use the same bare hands to wipe the sweat from their brows that they use to sort the thousands and thousands of old plastic bags.

American Corporations' Plastic Waste

Even though the women are working in a crowded slum outside Jakarta, Indonesia's largest city, much of the writing on the plastic garbage is in English. The women sort through liquid soap bottles, food wrappers, disposable diaper packages and huge bags with familiar logos—Dow, Du Pont, Monsanto, Solvay, BASF, Mobil. A white powder blows out of some of these bags as the women pull them from the pile. The women sorting the bags cannot read English, so they do not know that the white powder is titanium dioxide, which causes respiratory damage. They do know, however, that when the Indonesian plastics recycling companies they work for began importing plastic waste from the United States, they developed skin rashes they never had when they only processed locally produced plastic waste.

The health risks faced by the Indonesian women—and thousands like them throughout Asia and elsewhere in the Third World—are a direct result of the upsurge in plastic use in the United States—and of industry efforts to quell public concern in the United States about the environmental effects of increased

Anne Leonard, "Plastics: Trashing the Third World." This article first appeared in the June 1992 issue of *Multinational Monitor*, PO Box 19405, Washington, DC 20036; subscriptions, $22/year. Reprinted with permission.

plastics use.

In 1989, U.S. corporations used more than 12 billion pounds of plastic for packaging designed to be thrown away as soon as the package is opened. In the 1990s, this figure is expected to double.

It was not until recently, when people began to realize that landfilling plastic preserves it forever and burning it releases some of the most toxic substances known to science, that the U.S. public started to question the country's growing dependence on plastics. Rather than address these serious environmental problems, the plastics industry focused its attention on addressing its public image.

"Health risks faced by [Asian women] . . . are a direct result of the upsurge in plastic use in the United States."

A confidential December 22, 1989, letter from Larry Thomas, president of The Society of the Plastics Industry, invited plastics manufacturers to help develop a $150 million public relations campaign. "The image of plastics among consumers is deteriorating at an alarmingly fast pace. Opinion research experts tell us that it has plummeted so far and so fast, in fact, that we are approaching a 'point of no return,' " Thomas wrote. "Public opinion polls during the 1980s show that an increasing percentage of the general public believes plastics are harmful to health and the environment. That percentage rose sharply from 56 percent in 1988 to 72 percent in 1989. At this point we will soon reach a point from which it will be impossible to recover our credibility. (Witness what has happened to the nuclear energy industry.)"

Plastic's New Image

The plastics industry developed a two-point plan to restore its image. First, by mixing small amounts of corn starch into plastic products, the industry claimed its plastic packaging, garbage bags and diapers were "biodegradable." It did not take long for the U.S. public to figure out that although corn biodegrades, plastic does not.

Next, the industry jumped aboard the recycling bandwagon. Instead of "biodegradable," nearly every plastic package on the supermarket shelf is now stamped "recyclable."

"If we can get our act together and show the world just how recyclable these valuable polymers are and that the industry stands behind the commitment to prove it, then the mathematics will change," explained Marty Forman, chair of the Institute of Scrap Recycling Industries' plastic committee in 1991. "It won't be a 60-billion-pound market shrinking to 45 because 15 billion pounds were recycled, it will be an 80- or 90- or 100-billion-pound market which has expanded because those plastics are recycled."

Unfortunately, the plastics companies' claims that their plastic is "recyclable" are badly misleading. Plastic waste is seldom if ever recycled into the same product, so recycling used plastic does not make a dent in the amount of plastic needed to make the original products. Additionally, each time plastic is heated, its chemical composition changes and its quality decreases, so the number of times it can be recycled is very limited. The most dishonest aspect of plastic recycling claims, however, may be that many of the plastic bags and bottles dropped off at local recycling centers in the United States are shipped to Indonesia and other Third World countries, where much of it is not recycled at all.

Plastic Waste Exports

The plastics industry is now adopting the tried-and-true practices of international waste traders worldwide. By exporting their wastes to less-industrialized countries, U.S. plastics corporations have learned they can avoid domestic regulations and community opposition to waste-handling facilities, and pay their workers wages far below U.S. levels.

It is increasingly likely that the plastic bags and bottles dropped off at a local recycling center in the United States will end up in the countryside in China or in an illegal waste importer's shop in Manila.

In 1991 alone, over 200 million pounds of plastic waste were exported from the United States, according to data from Port Import/Export Research Service. This waste was sent to Argentina, Brazil, Chile, China, the Dominican Republic, Ghana, Ecuador, Guatemala, Hong Kong, Hungary, India, Indonesia, Israel, Jamaica, South Korea, Malaysia, Morocco, Nigeria, Pakistan, the Philippines, Russia, Singapore, South Africa, Taiwan, Tanzania, Thailand, and Trinidad and Tobago.

"In 1991 alone, over 200 million pounds of plastic waste were exported from the United States."

The primary target of U.S. plastic waste exporters is Asia. In 1991, more than 15 million pounds of plastic waste were shipped to the Philippines, 35 million pounds to Indonesia and over 75 million pounds to Hong Kong (much of which was sent on to China).

Industry recycling coalitions tout exports for diverting waste from diminishing U.S. landfill capacity while providing much needed employment in less-industrialized countries. In a September 1991 issue of *Plastics News*, Gretchen Brewer, a consultant with Earth Circle in La Jolla, California, justified plastic waste exports to Asia because "they have an urgent need to employ a lot of people, and it also helps them get more raw materials."

The U.S. Chamber of Commerce also denies that there are any problems with plastic waste exports. Harvey Alter, manager of the Chamber's Resources Policy Department, testified in 1991 in a congressional hearing on the subject.

"There is no basis," he assured lawmakers, "for accusations that the United States is 'dumping' hazardous (or other waste) on unsuspecting developing countries. Materials for recycling, virtually by definition, are sold to enterprises in countries with sophisticated manufacturing facilities."

Since there are no federal oversight mechanisms or standards for plastic waste exports, no one really knows what happens to the millions of pounds exported annually. Harrie Cohen, chief executive officer of Ontario Plastics Recycling in California, admits that he sends all of the plastic collected by his firm to China. "I don't know exactly what they're doing with it," he told a *Plastics News* reporter in 1991. Apparently, the U.S. "cradle to grave" approach to waste management, which requires tracking and monitoring at all stages from waste production to transport to disposal, does not apply if the grave is in another country.

A Greenpeace investigation of "recycling" facilities in Asia reveals that plastic waste is being shipped to countries which ban waste imports, that recycling facilities are endangering workers and the surrounding environment, and that much of the plastic sent to be recycled is simply dumped in landfills or in random locations.

U.S. plastics "recyclers" sent over 35 million pounds of plastic waste to Indonesia in 1991. The majority of the waste was sent to two cities on the island of Java—Jakarta and Surabaya.

Following Plastic Waste

Once the waste arrives in Indonesia, it undergoes labor-intensive sorting by hand. First, workers separate for disposal the non-plastic wastes—newspapers, clothing scraps, metal scraps, and miscellaneous other trash—that are imported along with the plastic cargo. Plastics that are either too contaminated or of such poor quality that recycling is not feasible are added to the discard pile. The owner of one Indonesian plastics recycling company estimates that up to 40 percent of the imported waste is directly landfilled at a local dump.

A large grinding machine transforms the separated plastic destined for recycling into flakes or pellets. Workers rinse these pellets with water to remove the residues and contaminants from the waste. This water, after rinsing many loads, is dumped onto the dirt floor or out the back door of the recycling plant.

> ***"Much of the plastic sent to be recycled is simply dumped in landfills or in random locations."***

The pellets, sometimes mixed with new plastic or other additives, are melted and forced through an extruder which shapes the hot plastic into long cords. These cords, once cooled, are chipped again and sent to manufacturing plants in Asia to be made into shoe soles, containers, or toys.

The plastic processing all happens indoors, in hot, crowded rooms with no ventilation systems. Recycling facilities in the United States, in contrast, are equipped with vacuum vents over the plastic melting machines to immediately remove fumes from the workers' environment.

Hong Kong is the largest single importer of U.S. plastic waste. In 1991, it received more than half of all U.S. plastics waste exports.

> ***"Underfunded customs and environmental agencies are unable to detect and intercept incoming waste shipments."***

The recycling facilities in Hong Kong look much the same as those in Indonesia. The same plastic grinding, melting, and re-shaping machines are in use, releasing the same strong noxious fumes. As in Indonesia, the untreated rinse water is discharged down drains or out the window. One difference is that, in real-estate scarce Hong Kong, the recycling shops are located on the fourth, eighth or eleventh floors of run-down industrial skyscrapers. The piles of plastics and the heat from the melting machines pose an obvious and frightening fire hazard in these densely packed buildings.

Waste in China

Of the 10 plastic recycling facilities Greenpeace investigated, the sophisticated equipment described by the U.S. Chamber of Commerce's Alter was nowhere to be seen. In fact, Hong Kong is only now in the beginning stages of negotiations for the first chemical waste treatment facility in the entire territory.

A number of plastic waste importers in Hong Kong simply warehouse waste en route to China. Many of these importers once processed the waste in Hong Kong but have since found land and labor costs much lower in the mainland.

A visit to a plastics recycling company in China revealed more of the same basic processes, but a different setting. Instead of the crowded slums of Jakarta or the industrial skyscrapers of Hong Kong, the facility was located on a dirt road in the countryside in Guangzhou. Since the area was suffering an electricity black-out the day of the investigation, none of the machinery was in operation.

The plastic waste is sorted in a walled-in courtyard which is also the site for worker housing. On one side of the courtyard, men pile huge cardboard boxes of plastic scrap. The boxes bear many of the same names as the plastic bags exported to Indonesia: Dow, Monsanto, General Electric, Du Pont. On the other side of the courtyard, less than 10 feet from the piles of plastic garbage, children play and women hang laundry to dry.

A massive pile of discards—unrecyclable plastics, clothing, scraps, and other garbage—occupy the center of the courtyard. The facility manager explains that there is no central dump in which this material can be disposed, so it is dumped in random locations in the countryside.

In August 1991, six containers exported from New York Harbor and supposedly carrying plastic waste arrived in Shanghai, China. When the Chinese plastic waste importers opened the containers, they discovered a grisly concoction of U.S. waste. Eight months later, the Shanghai City Environmental Protection Bureau sampled the containers and reported: "55 percent are . . . mostly household garbage, blood transfusion bags, and tubes. In order to prevent pollution, you must immediately request a professional unit to thoroughly sterilize the waste plastic and household garbage."

Although the Philippines has a strict law banning waste imports, U.S. corporations and waste brokers shipped over 15 million pounds of plastic waste to the country in 1991. As is the case in many waste-importing countries, underfunded customs and environmental agencies are unable to detect and intercept incoming waste shipments. And since the shipments are arranged covertly, the locations of the importing companies are unknown.

But the situation is not hopeless. Activists in the Philippines, working in conjunction with their counterparts in the United States, have begun to take steps to stop illegal plastic exports to the Philippines.

Environmentalists Take Action

Environmental and development organizations in the Philippines and the United States have formed a coalition to stop the illegal waste imports into the Philippines. Rene Salazar, director of the Southeast Asia Regional Institute for Community Education (SEARICE), a Philippine development organization and member of the new Coalition Against Toxic Waste, is demanding that his government take a more active role on this issue. "The toxic waste trade is a fast-growing industry as North Americans and Europeans do not wish to destroy their own backyards," he explains. "Mafia-like export companies are enticing Third World countries with potential profits to be made from trade in toxics. We have a complete list of all the imports of waste into the Philippines in 1991. We challenge the Philippine government to tell us where it went."

Using data from the U.S. Department of Customs, the new Coalition has begun tracking ships known to have carried the waste from the United States in the past. On April 1, 1992, activists from the United States and the Philippines boarded two waste-trading ships in New York Harbor: the Evergreen line's *Ever Guest*, which carried plastic waste to Indonesia, Hong Kong/China, and the Philippines, and the Mitsui OSK line's *Alligator Liberty*, which also carried plastic waste to Hong Kong/China.

> ***"Hopefully, the efforts of the Coalition will work to prevent waste traders from conducting 'business as usual.'"***

Wearing "Hazardous Exports Prevention Patrol" uniforms, the activists met with the captains of both ships and requested that they refuse to participate in the international waste trade. Maximo Kalaw, Jr., president of GreenForum Philippines, explained to the captains that, although the United States allows waste exports, unloading the waste in the Philippines is illegal.

After the captains refused to cease carrying waste from the United States to the Philippines or elsewhere in Asia, the Hazardous Exports Prevention Patrol hung a large banner on the *Ever Guest*'s hull which announced, "Hazardous Exports to Asia Begin Here." Philippine members of the Coalition Against Toxic Waste planned to greet the ships when they arrived in Asia.

Hopefully, the efforts of the Coalition will work to prevent waste traders from conducting "business as usual," says Nicanor Perlas, president of the Philippine Center for Alternative Development Initiatives (CADI) and a member of the Coalition. "We want to send a strong message to these irresponsible companies that the world has changed," Perlas says. "Citizens all over the world are informing each other and collaborating on ways to cut down the anti-social and environmentally destructive behavior of corporations."

Large Corporations Are Reducing Pollution

by John C. Newman

About the author: *John C. Newman is a senior vice president for Booz, Allen & Hamilton, a management and technology consulting firm in Bethesda, Maryland.*

Over the past two decades, corporations largely have viewed environmental compliance as another cost of doing business. The consequences of being fined, sued, or boycotted provided the primary incentive for companies to comply with environmental regulations and to take measures to reduce the ecological impact of their products and processes. During the past few years, corporations dramatically have shifted attitudes about the environment. Seeing carrots as well as sticks, they are modifying products, packaging, and practices to appeal to a growing market of environmentally conscious consumers.

An Environmental Commitment

Regulation still is an important motivator in industries ranging from automotive to energy. Yet, in many cases today—whether from fear of consumers, courts, or activists, or from enlightened self-interest—many corporations are *ahead* of environmental regulation. Companies large and small, in businesses as diverse as chemicals, consumer goods, and restaurants, are reducing toxic emissions, reformulating products, and redesigning packages to be environmentally friendly, beyond the letter of the law. Anyone who doubts this dramatic shift need look no further than the pages of corporations' annual reports:

• Eastman Kodak devoted a special four-page section to environmental initiatives, discussing efforts ranging from using recycled paperboard in its yellow film boxes to developing cleaner water-based chemistry for photo processing solutions.

• Phillips Petroleum featured environmental issues prominently in the headlines, text, and photos of two sections: Technology, covering subjects such as

John C. Newman, "Management Faces the Environmental Era." Reprinted, with permission, from *USA Today* magazine, November 1992. Copyright 1992 by the Society for the Advancement of Education.

reformulated gasolines, reduced emissions from production facilities, and modified refineries to produce low-sulfur diesel fuel; and Corporate Responsibility, stressing recycling efforts and elimination of polychlorinated biphenyls (PCBs) from plant sites.

• Borden devoted a special section to safeguarding the environment, citing the firm's efforts to design new food-service plants that minimize washwater and recover potato-starch (turning a former waste stream into a profitable product) and to use and mark food packaging to encourage recycling.

> ***"More than 80% [of senior managers] indicated environmental issues are 'extremely important.'"***

Monsanto, Exxon, and 3M, among others, also gave space in their annual reports to environmental concerns. In fact, in most corporate annual reports, it is unusual for the environment *not* to be mentioned. However, many ecological programs remain negatively driven; costs of environmental compliance are skyrocketing; and there is general disagreement on how to move effectively from a reactive to proactive mode in environmental management.

A High Priority for Managers

In 1991, Booz, Allen & Hamilton conducted a survey of 220 senior managers in major corporations to gauge their opinions and actions related to environmental management. The shift in opinion so evident in the marketplace was confirmed in the executive suite: more than 80% indicated environmental issues are "extremely important" to their company today. Just two years earlier, less than one-third made that statement.

Despite the numbers that cite environmental issues as extremely important, only a handful (seven percent) claim to be "very comfortable" that the environmental risks their company faces are well understood and that a comprehensive management strategy is in place to deal with them. This small group of leaders integrate environmental concerns throughout the company, manage their environmental policies higher and more centrally in the organization, and visibly demonstrate their ecological achievements to the marketplace and community.

Virtually all executives responding to the survey said they must manage risk better in the near future. Because the environmental function historically has been dealt with on a reactive basis (*e.g.*, figuring out how to comply with regulations), moving toward a more innovative environmental management posture that supports strategic objectives requires a major shift in emphasis and the direct involvement of top management. The corporations represented are large (91% have sales over $500,000,000, with 36% over $5,000,000,000); global (84% have plants in more than five countries); and diverse—consumer goods, chemicals, electronics, automotive services, natural resources, food and beverage, pharmaceuticals, and manufacturing predominated, with more than 15 different industries included. Most had formal environmental programs and published policies, but these efforts were relatively new. Seventy-five percent had established them since 1989.

Motivating Factors

"Negative" concerns—such as avoiding fines, lawsuits, or adverse publicity—motivate such policy development in most companies. The leaders (the seven percent minority), by contrast, say their programs are driven by opportunities that environmental trends can create for competitive advantage, and much less by bad press and lawsuits.

Not only do companies usually focus their environmental programs on negative concerns, they tend to be reactive in nature, concentrating on achieving compliance, cleaning up abandoned sites, and minimizing known liabilities. Most have a hard time dealing with the longer-term strategic issues that the environmental movement is creating. Nevertheless, respondents recognize the need for future improvement and say they are heading in the right direction.

Respondents believe that all major business activities are affected by environmental issues, with operations (manufacturing and distribution) being influenced most greatly. This would be expected because operating measures—waste generation, disposal, and minimization; energy reduction; recycling; compliance; and permitting—have been the focus of regulators. Reacting to this high level of impact, 90% say their corporations have incorporated formal mechanisms to consider environmental issues in manufacturing and distribution.

"Many managers see important parallels between quality control and sound environmental management."

Environmental management is expensive, disproportionately so in some industries. The surveyed companies indicate they spend an average of two percent of sales on environmental expenditures. Effective programs cut the cost of compliance and benefit the bottom line through reduced operating outlays, improved operations effectiveness, lower cost for waste disposal, and smaller expenditures for advertising and public relations because of their improved reputation.

Many managers see important parallels between quality control and sound environmental management. A key part of this orientation is lowering environmental costs throughout the product life cycle. These range from waste management and disposal expenses related to manufacturing and assembly to those involved in product use or disposal of packaging.

Bayer, the German chemicals company, expects to lower the amount it spends on environmental protection (today approximately 20% of manufacturing costs) by adopting an environmental quality orientation throughout the production process—for example, by instituting quality controls on raw materials and adopting waste minimization programs. Procter & Gamble has created concentrated detergents to reduce packaging; BMW is building a special plant for the design and manufacture of a recyclable car; and New England Electric has cut demand growth for electricity by one-third through aggressive conservation efforts.

Identifying Profit Opportunities

In addition to reducing costs, companies see profitable opportunities to launch ecologically sound products. Many are redesigning lines to offer goods with lower environmental impact, taking share from competitors' products with higher such expenses.

A number of surveys show that environmental concerns have become a significant factor in buying decisions. While traditional characteristics such as price and convenience are most important, a majority stated they would be more likely to purchase environmentally friendly products, such as a less-polluting formula of gasoline.

Beyond redesigning current products, significant business opportunities exist outside of many companies' core product lines. Solid and hazardous waste management and air and water pollution control technologies have an average growth rate of 21% in the U.S. alone. Companies in the energy, chemicals, and architecture and engineering industries have recognized this trend. Amoco, Du Pont, Exxon, Rhone Poulenc, Sandoz, Stone & Webster, and CRS Sirrine, to name a few, actively are pursuing environmental services ventures. The opening of Eastern Europe and growing concerns about ecological degradation in the developing world promise to increase this market exponentially.

The proliferation of environmental themes in companies' annual reports is no accident. As evidenced by the fast food industry (McDonald's, in particular), the environment has been *the* theme for 1990s advertising. *Advertising Age* now carries a regular column on "Green Marketing." Special events, newsletters, and media exposure on ecological themes have proliferated.

Since public opinion plays such a strong role in shaping environmental opportunities, corporations need to develop and execute a publicity strategy to promote their products and services and mitigate risk concerns. Those that establish a sound communications and outreach program stand to gain improved relations with consumers, regulators, environmentalists, and the public at large. For example, these relationships could facilitate the siting or permitting of a new facility, normally a multi-year process that can be stalled or stopped by public opposition. Finally, being perceived as environmentally acceptable or conscious improves

"The proliferation of environmental themes in companies' annual reports is no accident."

relations with buyers, increasing sales.

Nowhere is the importance of effective communications better understood than in the Chemical Manufacturers Association (CMA). Its Responsible Care initiative stresses the importance of two-way communications between corporations and customers, plants and communities. Responsible Care is backed by multi-million [dollar] advertising campaigns among chemical companies and CMA.

"An environmental program will succeed only if responsibility is ingrained in the corporate culture."

Clearly, publicizing of environmental achievement is a two-edged sword. Unless corporations have effective environmental management programs to back them up, public relations will be nothing more than window dressing. Today's consumers and environmental activists are far too savvy to be fooled. Just as risky as breaking environmental laws is being caught with empty claims. As manufacturers of disposable diapers, garbage bags, and other products can attest, even sincere efforts to make goods more ecologically friendly can be held up to intense scrutiny.

Here to Stay

No passing trend, the environmental era is here to stay. However, the novelty value of such activities and claims likely will fade as ecological management becomes an integral and expected part of doing business. Companies that succeed over time will be those establishing effective programs which promote a shift from opportunistic to strategic management of environmental issues.

There is no magic formula. Every company has unique products, processes, competitive position, and corporate culture that must be taken into consideration. However, the basic structure and benefits of a successful corporate environmental program are consistent. Because the opportunities and risks are moving targets, companies constantly need to strive to improve the effectiveness of their environmental programs to reap the rewards.

Two corporate mandates emerged from Booz, Allen's analysis of the survey: companies need to integrate ecological issues into total business operations and better manage their exposure to risk; and they must capitalize better on environmentally driven opportunities, both within their basic product lines and, where it builds from core business strengths, in new segments in the broader environmental services industry.

Environmental programs often are compared to total quality management, which relies on integration of that philosophy across all levels of the corporation. Similarly, environmental management must become a valued belief within the corporate culture. This largely will determine the success or failure of such a program.

The critical components of a comprehensive environmental management program include senior-level support and resource commitment; information management and communication; regulatory knowledge and tracking; organizational links between environmental management and business operations; a decision-making and planning process that addresses environmental issues; employee awareness and training; risk assessment and management; and an opportunity assessment and action plan.

Different Corporate Priorities

Some of these issues are being addressed by companies today, but many have not been successful in making major environmental improvements across the entire organization. Often, this is because these issues are handled in different ways, with little sharing of ideas or adoption of uniform standards to encourage excellence.

Environmental management can be viewed in a similar manner to financial management. Some companies include extensive checks and balances and encourage the highest ethical standards in financial matters. Others establish different priorities—short-term profits, new product deadlines—and tolerate less rigor in compliance matters. If a corporate culture allows spotty ecological compliance, it invites substantial environmental risk. Ultimately, an environmental program will succeed only if responsibility is ingrained in the corporate culture and championed from the top down, so individuals—from board members to plant operators—feel accountable for their actions.

Chemical Companies Are Reducing Pollution

by Frances Cairncross

About the author: *Frances Cairncross, the environmental editor of* The Economist *magazine since 1989, is the author of* Costing the Earth: The Challenge for Governments, the Opportunities for Business, *from which this viewpoint is excerpted.*

Companies face many pressures to think seriously about the environment. Most will resent them. A wise minority will see the opportunities they present and think creatively about ways to respond. They will make two kinds of changes: in their manufacturing processes and in the products they make. . . .

A Radical Change

Given the rising costs of being dirty, more companies see the benefits of being clean. The traditional approach has been "end-of-pipe" solutions, approaches that tackle effluents or gases just before, or even after, they leave the plant. But a new approach is now being developed: one of preventing pollution in the first place. It is less expensive in the long run to rethink the whole of an industrial process than to tack on a bit of extra technology at the end. As a result, "waste minimization" has become a catch-phrase among the greenest companies, whose names are rarely found on labels on supermarket shelves. Of the 100 to 200 companies worldwide that have made environmental performance their top concern, most are chemical companies. The most radical corporate thinking on the environment is taking place in large chemical companies such as Du Pont, Monsanto, Dow, Hoechst, ICI, and Ciba-Geigy. Realizing that the environment is one of the three or four most important issues facing their industry, their boards have formulated green strategies and set up sophisticated management systems to carry them out. This is hardly surprising: chemical companies in industrial countries typically produce 50% to 70% of all hazardous waste, either in the course of manufacturing or in the form of their final product. In other industries—especially oil and cars—there are companies that take greenery equally seriously, although they are rarer. But the number of companies and industries in the former category will rise as the costs of polluting increase.

These companies have tended to be more interested in making the manufacturing process cleaner than in producing goods for green consumers. They have focused mainly on the pollution that comes from smokestacks and sewage outflows. They have been driven much less by the buying power of shoppers than by the many costs of polluting, especially by the risks of handling and the costs of disposing of toxic waste. Many also claim that being greener has saved them money. Now, just as companies driven mainly by green consumerism have widened their attention from cleaner products to cleaner processes, so these businesses have begun to think more about their products.

Voluntary Efforts Exceed Regulation

A striking aspect of these environmental strategies is the extent to which they move far ahead of local regulatory requirements. For example, a number of companies have set themselves targets for toxic emissions and for waste generation far more stringent than anything the law requires. Monsanto pledged itself to cut toxic air emissions by 90% by the end of 1992 and then to work toward a goal of zero emissions. Du Pont promised to cut toxic air emissions by 60% from 1987 levels by 1993, to cut carcinogens by a further 90% before the end of the century, and eventually to stop emitting them entirely. These "green leaders" also frequently set standards for their overseas subsidiaries that may be even higher when measured against local norms. Frequently they insist that the company's standards be applied worldwide. For example, Union Carbide stipulates that its facilities in Africa stick to standards consistent with America's Clean Water Act, even though there is no such act in Africa. Johnson & Johnson applies the same standards all over the world. If standards rise in one country, the company claims, they are adopted universally. Dow Chemical has technology centers for various products, part of whose job is to ensure that the same technological standards are applied wherever a new plant is built. One effect is to raise standards in third world countries.

> ***"The most radical corporate thinking on the environment is taking place in large chemical companies."***

These policies have their critics. Few economists regard the goal of zero emissions as a wise one. Better to aim for the point at which the cost of getting rid of an extra molecule of nasty substance overtakes a rational estimate of the benefits to human health or to the environment. That point might come after a few puffs of carcinogens, especially from a plant in a densely populated area. It might be reasonable to set a much higher level of emissions in the case of less harmful gases, especially from a plant in the middle of nowhere. Some industrialists, even in the chemical industry, think that zero emissions are fine as an ideal but ludicrous as a practical goal. Robin Paul, managing director of Albright & Wilson, a British-based, American-owned chemicals-to-household goods group, is one such person: "Science," he points out, "is getting better all the time at measuring traces of substances in emissions." The chairman of Du Pont, Ed Woolard, made a broader point in a speech in December 1989:

> As we move closer to zero, the economic cost which society must ultimately bear may be very high. Or the energy expenditure necessary to eliminate a given emission may have more ecological impact than trade emissions themselves. Society will have to decide where the balance should be struck, and may conclude in some cases that zero emissions is neither in the environment's nor the public's best interest.

Imposing world standards has other drawbacks, especially in third world countries, where it may be harder than simply aiming to be a bit cleaner than the locals. It may indeed sometimes be worse for third world countries than playing by local rules. Excessive virtue may lead Western companies to invest less than they might otherwise do, leaving the door open to local firms more interested in evading than exceeding their country's green standards. A company's partner may be the government, which may balk at paying for higher emission standards than its own laws require. Or there may simply be no local facilities. IBM found no site in Argentina able to meet its tough requirements for waste disposal. Its Argentine plants therefore recycle three-quarters of the waste they generate.

> ***"Companies have set themselves [stringent] targets for toxic emissions and for waste generation."***

Pollution Control Costs

Pollution control still mostly means adding bits and pieces. It may mean installing a dust filter or building a purification plant, essentially transforming one type of waste into another that is less harmful and more manageable. For instance, a dust filter may convert uncontrolled clouds of filthy smoke into clean smoke and a heap of dirt that can be disposed of in a properly run landfill. Such an approach involves no change in the manufacturing process.

For companies in the industries most sensitive to the charge of polluting, add-on technology is not enough; the costs of being dirty and the benefits of being clean are so high that a new approach becomes worthwhile. The sums being spent on pollution control, especially in the chemical industry, are staggering. Bayer, a German chemical group, spends 20% of its manufacturing costs on environmental protection, about the same as it does on energy or labor. In the

United States, Chevron expects environmental spending to grow by 10% a year, and sees it glumly as "the only growth area of the oil industry." Albright & Wilson spends half its capital program on environmental protection projects or products. As a result, these companies find that it pays to take a radical approach to environmental protection. They have been seeking ways to ensure that, in the phrase coined by 3M, Pollution Prevention Pays (PPP). Proving its point, 3M claims to have saved well over $482 million in the 15 years that its PPP policy has been in effect. Other companies have acronyms of their own: Chevron has SMART (Save Money and Reduce Toxics), Texaco has WOW (Wipe Out Waste), and Dow Chemical has WRAP (Waste Reduction Always Pays).

"Frequently [green leaders] insist that the company's standards be applied worldwide."

Often simple improvements in process efficiency are looked for at first. Chris Hampson, board director of ICI responsible for the environment, says that about a quarter of the company's current environmental costs came from "losses in containment and less than optimum operation of plant." Robert Muirhead, formerly Exxon Chemicals' European safety and environmental control manager, reckons that waste-disposal costs are double the actual cost to the company when account is made of lost production and operating costs. When future liability for waste is added in, the cost could be doubled again.

Waste Minimization Programs

Having first popularized the idea of waste minimization, 3M now thinks through the concept in four stages, as follows.

1. *Reformulation.* Can a product be made with fewer raw materials so that the company warehouses are not full of dangerous substances? Can it be made with fewer toxic materials? For instance, can a solvent-based coating be replaced with a water-based one?

2. *Equipment redesign.* Can steam from one process, for instance, be used to drive another?

3. *Process modification.* Is it helpful to change, say, from batch feeding, which may mean readjusting the pollution-control systems with each new batch, to continuous feeding, with fewer quality-control problems?

4. *Resource recovery.* Can a waste product be salvaged and reused—as a raw material in another process or as a fuel or as a product?

Union Carbide breaks down the pattern in a different way. It is one of several companies that insist that each plant draw up its own waste-minimization policy. Since 1987, every big capital-investment project at Union Carbide has been reviewed for its potential in reducing waste. The company reckons that most of the measures it takes to minimize waste fall into one of three categories: good housekeeping, which includes reductions in spillages and leaks as well as better inventory control so that, for instance, chemicals are bought in smaller quantities; changes in the materials used, for example, switching to less hazardous substitutes; and changes in technology.

Good Housekeeping

The simplest waste-minimization techniques are often of the good-housekeeping variety. Stopping day-to-day accidents may not be as dramatic as preventing another Bhopal, but the cumulative effect on the environment may be as great. Some companies adopt techniques to stop accidents after they have had one. Sandoz, for example, Switzerland's second largest chemical company, spent Sfr 150 million ($100 million) on measures to prevent a repetition of the disastrous Schweizerhalle fire in 1986, including installing two catchment basins to stop water used in firefighting from draining into the Rhine. Shell UK spent £100,000 ($177,000) on a new leak-detection system after its spill in the Mersey in 1989. It is better, of course, to make such investments before accidents occur.

Better still is simple waste prevention. Union Carbide found that one of the best ways to stop the escape of polluted air from its plants was to set up a regular schedule of checks on components such as pumps, valves, and flanges. Keeping these promptly and properly repaired has allowed the chemical company to cut fugitive emissions down to a minute fraction of the levels judged acceptable by the Environmental Protection Agency. Utah-based Geneva Steel estimates it has been able to reduce its emissions to a quarter or a fifth of the allowable maximum primarily through the simple device of teaching workers to take proper care of the doors of its coke ovens. "The workers baby the things," says Joe Canon, its chairman.

Such good housekeeping is likely to be the least expensive kind of pollution prevention. Indeed, it is most likely to be the kind that shows a profit. "Most waste happens because something is not being used properly," argues ICI's Hampson. "Waste may happen because a plant is only 90% efficient in its use of raw materials. The rest is going out in waste. A lot of pollution is associated with inefficiency." As an example, he cites the production of fine denier nylon yarn by ICI's fibers division. The yarn is wound in a continuous thread several miles long onto a 25-pound bobbin. Each time the yarn snaps, ICI loses money—and creates waste, because the half-filled bobbin has to be disposed of. "Our efficiency is currently around 85%," Hampson says. "If we can get that up to the low 90s,

> ***"The sums being spent on pollution control, especially in the chemical industry, are staggering."***

we will make a 30% increase in profits."

Good housekeeping may stop waste being created in the first place. Beyond that, waste minimization becomes more complicated. It may involve changes in the quantity or quality of raw materials used in order to prevent waste being created in the first place. Thus Volvo is cutting solvent emissions from car plants by switching to water-based paints. Sigvard Höggren, vice president for environmental affairs, believes that "In the long run we must use materials in our processes that do not give rise at all to hazardous emissions." Polaroid is trying to find ways to substitute water for organic solvents in the manufacture of films for cameras. Exxon Chemicals, unhappy about the amount of hazardous waste caused by soil contamination, is rethinking the design of some of its plants. Often, simply using less air or water dramatically reduces the amount of polluted air or water that a plant has to dispose of. The volume of really nasty toxics may be no lower, but it may be easier to handle. Sandoz has been trying to reduce the amount of wastewater it generates. Handling fewer dangerous materials is one protection against accidents. Richard Mahoney, head of Monsanto, describes how his company reached that conclusion: "After Bhopal, we found we were handling far more hazardous materials than we needed to."...

"Together, wise government and inventive industry could be a formidable alliance for a greener world."

Government Intervention

Cleaning the environment requires government intervention. That can be done in both cost-effective and expensive ways. Some intervention is deeply harmful. Some is expensively unnecessary. But some is essential and has already made the industrial countries nicer places to live in than they might otherwise have been.

Companies must make sure that government intervention is conducted in ways that reward the cleanest and greenest. Industry's ingenuity can reduce enormously the costs of tackling environmental problems: by inventing substitutes for CFCs, by developing energy efficiency, by finding sustainable uses for rain-forest products, by reducing the rubbish heaps of the world, and by inventing simple, reliable forms of contraception. Harnessing that ingenuity by the skillful design of environmental policies is the challenge for governments. Together, wise government and inventive industry could be a formidable alliance for a greener world.

Petroleum Companies Work to Reduce Pollution

by American Petroleum Institute

About the author: *The American Petroleum Institute (API) is a Washington, D.C., trade association representing America's petroleum industry.*

Abraham Lincoln said, "Public opinion in this country is everything." That sentiment has even more relevance today. Americans are heavily influenced by issues directly affecting them: personal health and safety, and a clean environment. In fact, the 1990s have been labeled the Environmental Decade. This makes many Americans uneasy about the petroleum industry. They are skeptical that our industry can—or will—deliver on the issues that matter most to them. And we continue to wrestle with a fundamental dilemma—how best to continue providing a product that makes our American quality of life possible while, at the same time, achieving the heightened environmental expectations that have become part of basic American values. Over the years, the U.S. petroleum industry has worked to improve its environmental performance. Some of our progress has been triggered by public demands for changes and the resulting laws and regulations. That's how Americans traditionally have changed the status quo—and, fundamentally, how the American quality of life has evolved to be one of the best in the world: by exerting pressure on the forces that affect our lives, whether they are in the public or private domain. But past performance and progress do not adequately meet society's expectations today. The public asks more—greater environmental accountability and better health and safety protection for workers and communities, for example. Many Americans expect businesses to go beyond minimum requirements, especially in these areas.

Changing Public Perception

Americans believe the petroleum industry is driven solely by economic considerations—profits—and that this prevents its employees from sharing the val-

Excerpted from the American Petroleum Institute's 1993 booklet *STEP: Strategies for Today's Environmental Partnership*, pp. 2-3, 7-19. Reprinted with permission.

ues that ordinary citizens care about most. As a result, there is a gap between what people expect from the petroleum industry in terms of environmental performance and their perception of our industry's current performance. If we are to change how our industry is perceived, we must focus on two critical areas: "being a good energy provider and corporate citizen" and "demonstrating energy and environmental stewardship." To the public, this means—among other things—demonstrating that the industry is serious about protecting the environment. That's why API created STEP—Strategies for Today's Environmental Partnership. In 1989, the API Board of Directors identified seven key public policy trends of importance to the industry. From these trends, the Board selected the environment as its highest priority, and used it as the focal point of a new strategic planning effort, which is intended to provide increased long-range, action-oriented direction for our industry. The goal of STEP is environmental excellence through improved environmental, health and safety performance. We want to demonstrate to the public that we are dedicated to reducing the effects of our operations on the environment while economically developing the energy that is so important to our American way of life.

"Over the years, the U.S. petroleum industry has worked to improve its environmental performance."

Guiding Environmental Principles

An architect designs a building from the bottom up, all the while keeping the entire structure in mind. STEP is crafted in much this same way.

The foundation of our industry-wide effort to improve environmental, health and safety performance is API's Environmental Mission Statement and Guiding Environmental Principles, which were incorporated into API's bylaws in early 1990. Acceptance of the principles is a condition of membership in API. They serve as a long-term commitment to improved performance by API member companies.

The members of the American Petroleum Institute are dedicated to continuous efforts to improve the compatibility of our operations with the environment while economically developing energy resources and supplying high quality products and services to consumers. The members recognize the importance of efficiently meeting society's needs and our responsibility to work with the public, the government, and others to develop and to use natural resources in an environmentally sound manner while protecting the health and safety of our employees and the public. To meet these responsibilities, API members pledge to manage our businesses according to these principles:

• To recognize and to respond to community concerns about our raw materials, products and operations.

• To operate our plants and facilities, and to handle our raw materials and products in a manner that protects the environment, and the safety and health of our employees and the public.

• To make safety, health and environmental considerations a priority in our planning, and our development of new products and processes.

• To advise promptly appropriate officials, employees, customers and the public of information on significant industry-related safety, health and environmental hazards, and to recommend protective measures.

• To counsel customers, transporters and others in the safe use, transportation and disposal of our raw materials, products and waste materials.

• To economically develop and produce natural resources and to conserve those resources by using energy efficiently.

• To extend knowledge by conducting or supporting research on the safety, health and environmental effects of our raw materials, products, processes and waste materials.

• To commit to reduce overall emissions and waste generation.

• To work with others to resolve problems created by handling and disposal of hazardous substances from our operations.

"[API] members recognize the importance of efficiently meeting society's needs."

• To participate with government and others in creating responsible laws, regulations and standards to safeguard the community, workplace and environment.

• To promote these principles and practices by sharing experiences and offering assistance to others who produce, handle, use, transport or dispose of similar raw materials, petroleum products and wastes. . . .

Protecting Lives, Promoting Safety

Operating Hazards Strategic Plan. Protecting human lives, promoting safety and safeguarding the environment are central to the Operating Hazards Strategic Plan. The plan includes programs to systematically review, reevaluate and improve the way the petroleum industry does business. Programs under this strategic plan are designed to reduce crude oil and other product spills; improve safety at refineries; and get communities involved in programs with local petroleum plants. The industry recognizes that prevention is the first line of defense against oil spills. Therefore, it is setting and implementing higher training standards and developing solutions to potential problems. For example, an integrated electronic chart and navigation system for tankers that will improve the ease, speed and accuracy of critical decision-making on a vessel's bridge is being developed in collaboration with the U.S. Coast Guard. Such prevention mechanisms will be supported by the new Marine Spill Response Corporation, created and financed by the petroleum industry. When accidents happen, the MSRC will respond to catastrophic oil spills in coastal zone and tidal waters. Companies in the industry already belong to a variety of local oil spill clean-up cooperatives which deal with smaller incidents. The industry also has plans to reduce spills from other sources, such as tank barges on inland waterways and tank trucks on the nation's highways. Pipeline safety technology and training are being upgraded to prevent accidents. We expect these programs to reduce the likelihood of accidents on the job. API has published *Management of Process Hazards*, a comprehensive and rigorous standard that tailors safety measures to the distinctive processes of oil exploration, production, gas plant operations and refining. Guidelines, training seminars and workshops are being developed to help companies implement this standard. New technologies are also on the drawing board to improve plant safety. For example, to protect workers and the public from harmful levels of hydrogen sulfide that may result from crude oil production, API is conducting ongoing research on safety equipment and practices. . . .

Clearing the Air

Air Toxics Strategic Plan. The aim of API's Air Toxics Strategic Plan is to help reduce toxic emissions, while improving knowledge of their effects on human health and the environment. Many API members have underscored this commitment by participating in a new Environmental Protection Agency program for voluntary reductions in the emissions of 17 chemicals. About 2 percent of the 1.4 billion pounds of emissions targeted by EPA come from petroleum industry operations. API's Air Toxics Program includes initiatives to undertake a multiyear research program to improve air toxic measurement. It will:

- Establish an industry-wide database of emission trends.
- Perform research on the best ways to reduce emissions cost-effectively.
- Develop better methods to assess the exposure of workers and the public to industry emissions.
- Work with local communities to broaden communication and enhance understanding of industry emissions.

These initiatives are designed to further the broad goals of the Air Toxics Program, including improving the technical ability to measure and reduce emissions; advancing scientific understanding of risks to human health; and explaining to the public the nature and magnitude of any risks and what the industry is doing to reduce them.

"The industry recognizes that prevention is the first line of defense against oil spills."

Standards Review Program. API's standards program dates to the Institute's founding in 1919. Its purpose has been to help ensure safe industry practices in all its operations and to provide performance standards that can serve as the foundation for regulatory and legislative controls. For example, API standards describe how to install and maintain underground storage tanks, how to test for leaks, what materials tanks should be made of, how pipes should be connected, and so on. API currently produces and maintains about 400 voluntary standards through a collaborative effort of professional staff members and industry experts. All of these standards are based on sound technical and scientific principles, research and practice. And they cover all segments of the industry, including transportation, refining, marketing, measurement, safety and fire protection, exploration and production. To further safeguard human health and safety, and to help protect the environment, API committees will review and revise as necessary its current standards, recommended practices and guidelines, starting with some 90 that directly relate to the Guiding Environmental Principles. In addition, API is mounting initiatives that will lead to the development of 20 new standards. Some examples include:

> ***"API standards describe how to install and maintain underground storage tanks [and] how to test for leaks."***

• A recommended practice to ensure that dikes enclosing aboveground tanks are impermeable to spills.

• Additional release prevention design features for the bottoms of aboveground storage tanks.

• Training and certification of storage tank inspectors for aboveground and underground storage tanks.

• Guidance for abandonment of onshore oil wells.

• A recommended practice for the treatment and handling of petroleum marketing terminal effluent.

• A recommended practice for the improvement of contractor safety.

Tracking Progress

Environmental Performance Documentation Program. The Environmental Performance Documentation Program (EPDP) is designed to provide a public record of the industry's environmental performance. It will collect information on steps that the petroleum industry takes to promote environmental protection and safety in all sectors—and will measure, track, document and report on it. It will reflect the industry's efforts to assess and improve its environmental performance—a record that can be reviewed by Congress, government agencies and the public. The first phase of the documentation program will include analysis of:

• Data from the Toxics Release Inventory, which are collected by refineries to document the volume of chemicals that are released into the air, water or ground, or which are sent to offsite disposal facilities. These data are reported each year to the U.S. Environmental Protection Agency.

• Data on petroleum spills in U.S. navigable waters from offshore wells, pipelines, tankers and other non-industry sources. These data are reported each year to the U.S. Coast Guard.

• Data on occupational injuries and illnesses in the petroleum industry. These data, reported annually to the U.S. Occupational Safety and Health Administration, will help to demonstrate the safety of industry facilities and surrounding communities.

• Data on environmental expenditures by the petroleum industry. These data are broken down by sector—exploration and production, transportation, refining and marketing; by type of expense—whether capital, operating and maintenance, administrative, or research and development expenditures; and by pollutant type—air, water, solid waste, remediation, spill cleanup and others. . . .

Going to the Source

Pollution Prevention Program. The federal Pollution Prevention Act, which became law in late 1990, underscores the federal government's growing interest in keeping the environment clean by avoiding pollution in the first place. The law commits the U.S. Environmental Protection Agency (EPA) to developing a strategy to realize this goal. EPA plans to make pollution prevention a model of programs to protect the environment. Many of the nation's major petroleum companies have joined in EPA's voluntary 33/50 project, which aimed to cut emissions of 17 targeted toxic chemicals by 33 percent in 1992, with 50 percent cut by 1995. Even before EPA's 33/50 project was announced, members of the petroleum industry had taken voluntary steps to incorporate pollution prevention practices in their operations nationwide. Individual company efforts such as SMART (Save Money And Reduce Toxics) and WOW (Wipe Out Waste) have been adopted to promote and institutionalize waste management efforts. As the Council on Environmental Quality's 1990 annual report noted, SMART nearly halved the amount of hazardous waste disposed of by one API member company in just one year.

"Many of the nation's major petroleum companies . . . [aim] to cut emissions of 17 targeted toxic chemicals."

API's Pollution Prevention Task Force is spearheading the industry's commitment to identifying and implementing pollution prevention opportunities. It includes staff members who work on programs dealing with air, water and waste, as well as those who deal with exploration and production, transportation, refining and marketing programs. The task force has two main objectives:

• To enable API and its member companies to evaluate and respond to emerging pollution prevention issues.

• To identify research, emission and waste reduction opportunities and develop industry-wide practices, thereby actively leading industry pollution prevention initiatives.

Crisis Management Program. A crisis can be almost anything: It can take any form, affect any person, company, or aspect of our business, and occur at any time, even in the best-managed companies. Toward the end of 1989, API's General Committee on Communications (GCC), responding to the industry's need to explore and expand our ability to deal with crises of various types, launched a crisis management program. The GCC began an ongoing program of seminars to offer counsel from experts and to provide a forum for the exchange of new ideas to enhance our knowledge and understanding of crisis management. The Institute has assembled a crisis management team of senior executives. Their combined experience will serve as a valuable resource in helping companies manage various kinds of crises. API's 24-hour Crisis Hotline—1-800-673-2778—is operational, and companies are urged to notify API of any crisis events such as fires, explosions and oil spills.

Oil Spill Cleanup

On March 24, 1989, an oil tanker struck a reef in Alaska's Prince William Sound, precipitating the biggest oil spill in the history of the United States. Within three months, the industry announced it would create the world's largest oil spill response organization to combat catastrophic spills in U.S. coastal or tidal waters. In August 1990, the independent Marine Spill Response Corporation (MSRC) was born. MSRC will operate five regional response centers around the coastlines of the United States, each supported by a number of strategically placed equipment sites. The Marine Preservation Association, an organization including oil companies and other shippers or receivers of oil, funds MSRC but it has no control over its operations. MSRC, based in the nation's capital, will operate its five regional response centers in the New York-New Jersey metropolitan area; the Miami-Fort Lauderdale area of South Florida; Lake Charles, La.; Port Hueneme, Calif., north of Los Angeles; and Seattle, Wash. When it becomes fully operational, MSRC will provide a best-effort response to cleaning up catastrophic spills. Each of the regional centers will maintain state-of-the-art technologies to respond to oil spills as large as the Alaska spill. MSRC will execute the spiller's response plan under the direction of the U.S. Coast Guard, which has a presence around the nation's coasts and an effective command, control and communications structure. Coast Guard jurisdiction will not relieve oil spillers

> ***"Regional centers will maintain state-of-the-art technologies to respond to oil spills as large as the Alaska spill."***

of their responsibility for the cleanup. It will, however, provide clear direction and coordination of cleanup operations when needed, because the Coast Guard, by law, has the ultimate authority in dealing with oil spills. As part of its efforts to prevent spills before they happen, MSRC will administer a comprehensive research and development program to improve the knowledge and technology used to respond to and clean up spills. This program will complement other programs in government, academe and industry. MSRC studies will include those on preventing loss of oil from ships, on-water oil recovery and treatment, preventing and mitigating shoreline impacts, fate and effects of spilled crude oil and products, mitigating impacts on wildlife, and health and safety. MSRC will coordinate its efforts with an estimated 150 existing oil spill cooperatives and subcontractors. Over the years, these groups have compiled a solid record of handling non-catastrophic spills—which account for some 99.9 percent of all spills. MSRC is designed to complement this existing capability—that is, to handle the infrequent, yet potentially catastrophic-sized spills. Existing cooperatives and subcontractors would continue to handle smaller spills, except in the instances in which MSRC's assistance might be required or when the Coast Guard requests MSRC's aid. Members of the Marine Preservation Association will pay annual dues, based on the number of barrels of oil they transported in the previous year. MSRC will use these funds to pay its annual operating, capital, and research and development costs. . . .

"API and the industry have the structure for a long-term commitment to health, safety, and the environment."

Cleaner Cars, Cleaner Fuels

The gasoline-powered automobile has been around for a century, but until recently little research had been done to optimize fuel composition and modern vehicle emission control systems. The Auto/Oil Air Quality Improvement Research Program was launched in 1989 to help fill this knowledge gap. Sponsored by the Big Three domestic automakers and 14 oil companies, this program is the first in history to consider cars and trucks and the fuels they use as a total system. As the program was being created, advice was sought from the U.S. Environmental Protection Agency, California Air Resources Board, Northeast States for Coordinated Air Use Management, and other regulatory organizations. A group of academics, from schools and laboratories including the University of California at Berkeley, Princeton, Massachusetts Institute of Technology, California Institute of Technology, and the National Center for Atmospheric Research, was assembled to oversee the program's efforts. The testing program is being conducted round-the-clock in EPA-approved laboratories. More than 2,200 different tests using 29 different fuels and fuel blends are being conducted as part of the program. The varying types of gasoline were created by changing the concentrations of gasoline components such as aromatics, olefins and oxygen. Methanol blends, ethanol and natural gas are also being investigated during the testing program. These fuels are being tested in cars that have seen real use, as well as laboratory prototypes of alternative-fuel and other new vehicles. All types of vehicle emissions—exhaust, evaporative and running losses—are being measured to gain a better understanding of the role each of these emissions plays in smog formation. Results from these tests are being subjected to gas chromatographic analysis (an advanced way of measuring potential pollutants). The program will produce the largest number of such analyses ever carried out anywhere with respect to vehicle emissions. Data from the research program will be used in atmospheric chemistry and air quality models to determine the potential reductions in urban ozone and other pollutants, which may be achieved by reformulating gasoline and using alternative fuels. The findings from the initial round of testing have already been provided to the EPA and other organizations looking into ways of reducing automobile pollution. . . .

API and the industry have the structure for a long-term commitment to health, safety and the environment in place. Using the recommended management practices as flexible guidelines, member companies can construct their own individual programs. Through STEP—Strategies for Today's Environmental Partnership—we are working to encourage improved performance industry-wide.

Chapter 3

How Effective Is the Environmental Protection Agency?

Chapter Preface

For more than a decade, one of the top priorities of the Environmental Protection Agency (EPA) has been the cleanup of more than one thousand toxic waste sites. Specifically, the EPA uses the 1980 environmental act commonly known as Superfund—an amalgam of laws, agency personnel, and tax-derived funds—to force companies, often through lawsuits, to pay for toxic waste cleanup. For at the heart of Superfund is a strict "joint and several" liability provision, which declares that any company that produces, transports, or dumps toxic waste can be held liable for the entire cost of cleanup, no matter how little of the waste a company is responsible for.

Much of the debate concerning the EPA's effectiveness centers on this controversial provision. EPA critics, many corporate leaders among them, argue that the agency's pursuit of potentially responsible companies that are reluctant to pay generates lawsuits and litigation gridlock—cases that can drag on both in and out of courts for years—draining public and private funds. As President Bill Clinton stated in 1993: "Superfund has been a disaster. I would like to use [it] to clean up pollution for a change and not just to pay the lawyers."

Yet the legal wrangling ultimately succeeds in getting toxic waste removed, many environmentalists contend, and the prospect of such litigation spurs other companies to adopt environmentally sound policies. As William Roberts, legislative director for the Environmental Defense Fund, states, "We would like to preserve joint and several. It has great pollution prevention value because it forces polluters to pay." Jonathan Lash, president of the World Resources Institute environmental think tank, agrees and adds that the provision has "given birth to an enormous greening among corporate leaders. The enormous concern about environmental liability has led to a lot of voluntary cleanup and pollution prevention efforts."

Both critics and defenders of Superfund agree that completing toxic waste site cleanup is slow and arduous. The agency aims to speed this process, in part by agreeing to settle Superfund cases early with companies that are only marginally responsible for toxic waste. Much of private industry, however, calls for quicker cleanup through the reform or outright elimination of Superfund, and advocates that companies voluntarily clean toxic waste sites with EPA approval. The authors in this chapter discuss the effectiveness of Superfund and other EPA activities.

The EPA Is Effective at Reducing Hazardous Waste

by William K. Reilly

About the author: *William K. Reilly served as administrator of the EPA from 1989 to 1993. Reilly, a former president of the World Wildlife Fund, is now a senior fellow at the fund, an organization in Washington, D.C., that promotes wildlife preservation.*

When Superfund was created in 1980, few people understood the magnitude of the problem of uncontrolled hazardous waste sites the nation faced. Many believed that it would be a short-lived program of perhaps five years to clean up problems like [New York's] Love Canal. No one was prepared for the thousands of sites per year that were in fact reported to the [Environmental Protection] Agency. Ten years later, we know that the problem is much more complex and widespread.

Today, the Superfund program is one of the Agency's largest and most visible programs and has generated its share of controversies. In its early years, Superfund suffered from many problems, not the least of which was the lack of expertise in hazardous waste cleanup. Having made so much progress in understanding often complex sites and the technology needed to clean them up, we sometimes forget that we were pathfinders only ten years ago and that we are still breaking new ground in how to clean up wastes today.

Hazardous Waste Sites

Since 1980, the Agency has had in its inventory reports of close to 35,000 potential uncontrolled hazardous waste sites; has identified more than 1,200 sites for priority cleanup under Superfund; and has started remedial action at over 600 of these sites. More than 2,600 cleanup actions called "removals" have been performed under emergency authority, and immediate, acute threats have been eliminated at all National Priorities List (NPL) sites. (This is the list that

From William K. Reilly's statement to the House of Representatives' Committee on Public Works and Transportation, Subcommittee on Investigations and Oversight, October 3, 1991.

governs where we'll focus our long-term cleanup efforts.)

Although over one-third of the sites of the National Priorities List have had a removal action, permanent cleanup of an NPL site typically involves a more complex remedial action. Matching the capability of technology at the site to the nature of the waste, and balancing that match with the need to be protective and effective over the long term, as well as being cost effective, meeting applicable standards (and other legal requirements), and addressing State, local and citizen concerns requires technological sophistication and public sensitivity.

> ***"I believe we have made significant improvement in the consistency and quality of our Superfund program."***

In addition, most sites have a number of potentially responsible parties (PRPs). Sites with hundreds of responsible parties are not uncommon. The private resources provided by responsible parties—added to the resources of the Fund—are absolutely essential to cleaning up sites. The complex system for bringing responsible parties into the process itself has resulted in the addition of over $4 billion to the public expenditures at Superfund cleanups. Finally, layered on top of all of this are issues of overwhelming technical difficulty such as groundwater cleanup.

Superfund Accomplishments

When I testified before Congress during my confirmation hearings, I promised to undertake a comprehensive study of the Superfund program. That study—the "Superfund Management Review" or "90 Day Study"—was completed in 1989. Acting on its recommendations, we have made continual improvements to the Superfund program, and substantial progress toward cleaning up the worst of the nation's uncontrolled hazardous waste sites. We are particularly proud of our accomplishments in several key areas:

addressing immediate threats;

moving ahead on permanent remedies;

applying "enforcement first" principles; and

encouraging innovative technologies.

In each of these areas, I believe we have made significant improvement in the consistency and quality of our Superfund program. . . .

While the program is working effectively, it is not perfect. It draws criticism from many sides. That is to be expected, in part, because of the very nature of the program. After all, we are dealing with the very worst uncontrolled hazardous waste sites. These sites often are near residential communities, where people feel they have a direct stake in the Superfund site cleanup; we are often using advanced or innovative technologies that can be expensive and that push the frontiers of our scientific understanding; and private and governmental parties are finally forced to confront the disposal practices of years ago.

We in government have an obligation to protect public health and the environment, but must do so in a fiscally responsible manner. That is why I welcome the attention focused on this program, so that we can meet both objectives in the best manner possible.

Addressing Immediate Threats

The risk posed by Superfund sites to human and environmental health are very real and the Superfund program is committed to eliminating immediate threats posed by hazardous waste sites as soon as we find them. When a site is reported to us, either we at the EPA, or the State under our aegis, conduct an assessment to determine whether an emergency action is necessary to reduce imminent threats at the site, and to determine whether the site should be identified as a national priority and placed on the NPL.

Nearly 35,000 sites have been reported to EPA. So far 93 percent of these sites have been evaluated to ascertain the potential threats posed by the site. More than half of these have been determined to need no further Federal action and, where appropriate, have been referred to State governments for their attention. A total of 1245 sites (or about 4 percent) in the site inventory have been placed on the NPL, for potential longer term remedial measures under the Superfund program. In addition, we have taken emergency site stabilization and cleanup actions in over 2600 cases.

"The Superfund program is committed to eliminating immediate threats posed by hazardous waste sites."

Recognizing that long-term cleanup at NPL sites takes years, one of my initiatives was to assess proposed and final NPL sites to determine whether an immediate threat exists and to take emergency action where needed. As a result of these assessments, and our follow-up response actions, we are now confident that there are no immediate threats at NPL sites. We will continue to reassess at least half of all NPL sites each year to ensure that we can continually inform the American people that immediate threats have been eliminated at NPL sites.

Each Superfund site has its own unique characteristics, and cleanups must be tailored to the specific needs of each site and the types of wastes that contaminate it. For example, the physical aspect of the site, such as its hydrology, geology, topography and climate, determine how contaminants will affect the environment. Variations in site type (e.g. landfill, manufacturing plant, military base, and metal mine), the kinds of waste at the site, and the risk presented by the site, add to the complexity of remedy evaluation and selection. Since information on the health and environmental effects of hazardous substances comes mainly from laboratory studies of pure chemicals, there is still much to learn about the nature of the toxic "soups" generally found at these sites, how they affect the environment, and how best to control them.

Incremental Improvements

Since the enactment of the Superfund Amendments and Reauthorization Act (SARA) in late 1986, progress toward achieving complete remedies has increased dramatically. There has been greater than a threefold increase in number of sites where remedies have been selected and designs and construction projects have been initiated.

It is important to remember that many of these sites were decades in the making, and as much as we wish otherwise, their cleanup cannot always be quick and easy. An analogy can be made between Superfund actions and the cleanup of our rivers. Although sewage treatment plants are constructed across the country, the day the plant goes on line does not result in instantaneous cleanup of a river. It is the incremental improvements that occur over time that will provide us with a noticeably cleaner environment in the long term.

Superfund is achieving these kinds of incremental results all across the country. Construction activity, either under remedial or removal authorities, has occurred at more than half of the NPL sites. Let me cite some of the progress resulting from these activities:

1. All surface contamination (e.g. soils, buildings and debris) has been cleaned up at 196 sites.

2. All construction work has been completed at 63 sites. We expect to see a dramatic acceleration in the number of sites completed over the next few years.

About 10 percent of the U.S. population (23 million people) within 4 miles of an NPL site have benefitted from Superfund actions. Many more have benefitted from emergency response actions at non-NPL sites. Water safe for drinking and bathing has been provided to a total of 450,000 people—roughly the population of Atlanta, GA—at 185 sites. We have relocated out of harm's way more than 30,000 people—a city the size of Annapolis, MD.

> ***"About 10 percent of the U.S. population (23 million people) . . . have benefitted from Superfund actions."***

• Superfund has treated or properly disposed of enough contaminated soil and debris to cover a football field a mile high.

• It has treated and disposed of more than one billion gallons of liquid hazardous waste. This is equal to 4 gallons of liquid waste for every man, woman, and child in the United States.

• Finally, it has cleaned contaminated groundwater almost equal in amount to that consumed by New York City residents in a five year period.

These indicators, and others like them, must become the measures by which we gauge the effectiveness and impacts of our hazardous waste cleanup program.

Treatment has become an important component of our response actions. In fiscal year 1990, 79 percent of remedies selected employed treatment of some kind to eliminate hazardous substances (instead of relying solely on containment). This is a dramatic increase over fiscal year 1986, when less than 50 percent of the remedies selected included treatment measures.

"We get more sites cleaned up by combining resources of the responsible parties and the Fund."

Although we are treating wastes at more sites, treatment and control of contaminated groundwater is an area of substantial technical challenge and unfortunately, I do not see easy solutions. Pumping and treating groundwater reduces the volume of chemicals and retards the movement of contaminated plumes throughout the aquifer. Groundwater control or remediation is a lengthy process; it may take decades to cleanse contaminated aquifers. While the feasibility of restoring aquifers for drinking water use may still be uncertain, we have seen some promising results from our efforts to pump groundwater and treat contaminated groundwater. We continue to work with the academic and technical communities to develop better techniques to evaluate and remedy groundwater contamination.

Enforcement First Is Working

I would like to turn now to enforcement. The SARA legislation in late 1986 strengthened EPA's enforcement hand and underscored the importance of responsible party site cleanup. Responsible party participation in site remediation serves two goals.

First, assuming that most sites on the NPL have a good prospect for responsible party participation in cleanup, it dramatically expands the amount of effort being devoted to site cleanup, and thus accelerates the rate at which sites can be brought under control. Where responsible parties are not found, EPA uses the Fund. We get more sites cleaned up by combining resources of the responsible parties and the Fund.

Second, it heightens our nation's awareness of contamination of the environment and sends a clear signal about accountability, deterring future waste generation and promoting good hazardous waste management and site cleanup in anticipation of future property transfer.

Armed with congressionally mandated tools, and with the Superfund Management Review in place, the Agency began to move forward aggressively. By carrying out some 50 major recommendations in the Superfund Management

Review, including a significant infusion of resources into our legal and technical field enforcement staff, we have managed to make "enforcement first" a reality of which we can be justifiably proud.

The broad statistics tell a very positive story about the use of enforcement tools, and the results we have achieved. First, I will share with you the results. In 1989 and 1990, responsible party participation in site cleanup has climbed from less than a third of all projects to more than 60 percent. When you consider that at a number of sites are so-called "orphan" sites, with no easily identified responsible parties capable of carrying out cleanup, this enforcement achievement is even more notable. With respect to the value of work commitments, the recent history is equally positive. Since 1980 we've obtained commitments for over 4 billion dollars of work from responsible parties. More than half of this, 2 billion dollars, has been obtained since 1989. In short, enforcement has yielded a very substantial site cleanup dividend. . . .

Superfund: An Ambitious Program

Cleaning up the nation's worst hazardous waste sites is a monumental challenge. As we continue to discover new sites and clean up those already identified, it is clear that Superfund is one of the most ambitious environmental programs ever undertaken. Through this program, we are responsible for protecting human health and the environment from uncontrolled hazardous waste sites as quickly and as effectively as possible.

Superfund has truly been a learning experience over the past ten years. Congress has given us guidelines and mandates for how to implement the program, and I believe we have accomplished a great deal. Virtually all of the 35,000 sites in the inventory have been evaluated, and more than 1200 have been included on the National Priorities List. Immediate threats are under control at all National Priority List sites, and work is underway at most sites. We are successfully making responsible parties pay for cleanup and we are encouraging the development of promising new technologies.

I am proud of this program's accomplishments, though EPA cannot take all the credit. This has truly been a team effort with the Department of Justice, the Army Corps of Engineers, and with the States. Our sense of accomplishment is shared with the local governments and community groups at each of these sites. And finally, it is shared with the responsible parties, who in many cases have taken a forward position in implementing emergency and remedial actions.

But progress is not perfection. As we have learned more about the environmental impacts of uncontrolled hazardous waste sites and the technologies necessary to clean them up, we have encountered many problems. Some are technological, some are managerial; despite these problems, I believe we can effectively implement the existing statute.

The EPA Responds Effectively to Pollution Emergencies

by Environmental Protection Agency

About the author: *The Environmental Protection Agency (EPA), created in 1970, is the federal agency in charge of controlling and preventing air and water pollution caused by pesticides, radiation, and other toxic substances. EPA headquarters are in Washington, D.C.*

EPA's emergency response program is a part of the Agency's implementation of the Comprehensive Environmental Response, Compensation, and Liability Act of 1980 (CERCLA). Also known as Superfund, this law was amended by the Superfund Amendments and Reauthorization Act of 1986. CERCLA established a "Superfund" to finance cleanup of the worst hazardous waste sites and set criteria for emergency notification of releases of hazardous substances.

Under its Superfund program, EPA conducts remedial actions (longer-term cleanups) and removal actions (short-term responses). Emergency response to hazardous substance releases are removal actions.

Oil Spills

Emergency response to oil spills is authorized by the Clean Water Act, as amended by the Oil Pollution Act of 1990. The relationship between hazardous substance and oil spill response is reflected in the National Oil and Hazardous Substances Pollution Contingency Plan, a set of Federal regulations that integrates requirements for responses to oil spills with requirements for responses to hazardous substance releases.

EPA's emergency response program is managed by the Emergency Response Division within the Office of Emergency and Remedial Response at EPA Headquarters. However, actual emergency response actions are conducted by On-

From Environmental Protection Agency, *An Overview of the Emergency Response Program*, April 1992.

Scene Coordinators and other frontline staff based in the ten EPA Regional offices throughout the United States.

EPA is one of several Federal agencies that manage the response to environmental incidents. The U.S. Coast Guard, Department of Energy, and others play key roles in responding to oil spills, hazardous substance releases, and radiological events.

An EPA representative chairs the National Response Team, an organization of 15 Federal agency representatives which coordinate national spill response and preparedness.

The National Response Team provides support to the Regional Response Teams, which include Federal and State representatives from each of the Regions. The Regional Response Teams are co-chaired by EPA and Coast Guard representatives. These teams coordinate regional spill response planning and preparedness, as well as direct support to On-Scene Coordinators.

For releases of national significance, EPA activates its National Incident Coordination Team to efficiently mobilize the full resources of the Agency and ensure rapid transfer of information within EPA.

Responding to Emergencies

EPA's Environmental Response Team is a group of scientists and engineers that provides 24-hour technical expertise to On-Scene Coordinators, State and local responders, and other countries in times of international environmental crisis. The Team's on-site support can include sampling and analysis, assessment of hazards, and evaluation and implementation of cleanup techniques. The Environmental Response Team also trains over 5,000 people each year for emergency response operations. Training courses cover worker health and safety and other aspects of identifying, evaluating, and controlling hazardous substance releases.

EPA's emergency response program is rounded out by a network of cleanup contractors strategically placed throughout the country to provide immediate cleanup capabilities. Technical Assistance Teams conduct site investigations and sampling, handle the required record-keeping, and help the Agency provide information to the public. Actual site cleanups are managed by EPA and performed by companies contracted through the Agency's Emergency Response Cleanup Services.

"EPA is one of several Federal agencies that manage the response to environmental incidents."

Taken together, these efforts are designed to ensure prompt, safe, and effective EPA response to hazardous substance releases and oil spills. Regardless of who is responsible for cleanup and oversight, however, EPA maintains the capability and authority to provide a Federal presence at sites to ensure protection of public health and welfare and the environment.

Large-scale industrial production in the United States has generated vast quantities of hazardous substances. Prior to the 1970s, the U.S. paid little attention to the disposal of these industrial hazardous substances. As a result, hazardous substances often were buried in pits, dumped in ponds and lagoons, or mixed with non-hazardous waste in municipal landfills. The continuation of these practices over many years has resulted in the creation of tens of thousands of hazardous waste sites.

During the 1970s and early 1980s, Federal, State, and local government officials and citizens became more aware that uncontrolled hazardous waste disposal threatened public health and our finite natural resources. As a result, previously limited legal and regulatory controls over hazardous waste disposal were expanded to include a number of comprehensive laws and regulations. These efforts are designed to clean up the mistakes of the past, cope with the hazardous substance emergencies of the present, and create guidelines for prudent hazardous waste management and disposal for the future. . . .

Congress strengthened the foundation of the Superfund program in 1986 when it enacted the Superfund Amendments and Reauthorization Act (SARA). SARA provides the Federal government with increased enforcement powers against responsible parties to compel them to meet their legal obligations of reporting releases of hazardous substances and funding cleanup efforts.

"The Environmental Response Team also trains over 5,000 people each year for emergency response operations."

Congress also passed the Emergency Planning and Community Right-to-Know Act (EPCRA) in 1986 as Title III of SARA. SARA Title III expands the role of State and local governments and citizens in emergency planning processes and emphasizes the importance of emergency response planning and training programs.

The Superfund Trust Fund

The Superfund trust fund was created as an important component of CERCLA to give the Federal government flexibility in identifying and addressing potentially harmful releases of hazardous substances. The fund is derived from taxes on petroleum and the production of a number of commercial chemicals. The Superfund enables EPA and the Coast Guard to respond immediately to hazardous substance releases and contamination problems that pose a threat to public health and the environment. The Agency may seek repayment later from the party or parties responsible for the release.

The philosophy behind Superfund is that protecting public health and the environment from the risks of hazardous substance releases requires a timely response action; a delayed response can compound the problem and cost more money in the long run.

As soon as the person in charge of a facility or vessel knows that a hazardous substance has been released in a reportable quantity (RQ) or more into the environment, he or she must immediately notify the National Response Center (NRC). In addition, SARA Title III requires that the owner or operator of a facility report such releases to the appropriate State emergency response commission (SERC) and local emergency planning committee (LEPC).

> ***"Large-scale industrial production in the United States has generated vast quantities of hazardous substances."***

Notification of a release triggers the National Response System, a network of Federal, State, local, and private sector roles and responsibilities for responding to releases of hazardous substances.

When the National Response Center is notified, the duty officer relays the release information to an EPA or Coast Guard On-Scene Coordinator (OSC), depending on the location and nature of the release. After receiving a report of a hazardous substance release, the Federal OSC evaluates the risk based on the circumstances of the release. All notifications to the Federal government are subsequently collected in the automated Emergency Response Notification System (ERNS).

The Federal Role

The Federal response process begins with the OSC's decision to initiate response measures. This decision is based on a preliminary assessment of notification information and on follow-up data gathered from the site.

Once the OSC decides that the Federal government is the appropriate response agency, he or she must then decide what type of response to make: whether the release should be contained to prevent migration and treated in place, or whether off-site disposal or treatment should be undertaken.

When necessary, expertise of other Federal agencies can be brought to a response action through the National Response Team and the Regional Response Teams. The Federal OSC also may seek assistance from other States, the EPA Regional offices, or the Environmental Response Team. State agencies or the parties responsible for a release may take the lead in cleaning it up when the OSC determines that Federal assistance is not necessary. In these cases, the Federal OSC may retain the authority for oversight or monitoring of the operations to ensure that the threat is properly mitigated.

The National Contingency Plan provides the guidelines and procedures for responses to hazardous substances, and authorizes EPA to conduct two types of response activities: removal actions and remedial actions.

Removal actions are short-term, relatively low-cost actions that EPA can take in response to a release or a threat of a release of a hazardous substance that poses an immediate danger to public health or the environment. Remedial actions are longer-term efforts to select a remedy and more fully clean up the site. The Agency's emergency response program, the focus of this viewpoint, is responsible for carrying out removal actions.

Hazardous Situations

Examples of situations that might require a removal action include: fires or explosions, threat of human exposure to a hazardous substance, contamination of a drinking water supply, or releases from abandoned facilities. If a preliminary assessment of the situation shows that there is an immediate threat and that no other capable authority can respond in a timely manner, a removal action may be the appropriate solution to the problem.

> ***"The Superfund enables EPA and the Coast Guard to respond immediately to hazardous substance releases."***

Removal actions are limited by CERCLA to one year and $2 million. Although the National Contingency Plan reflects these time and spending limits, exemptions may be obtained when a continued response is required to prevent or mitigate an emergency, or when further response actions are appropriate and consistent with the remedial action to be taken.

Removal actions can involve many different activities to stabilize or eliminate the threat posed by a release, including:

- Excavating or pumping hazardous substances for treatment or disposal offsite;
- Providing alternate water supplies;
- Treating hazardous substances on-site;
- Relocating residents temporarily; and
- Installing fences to reduce the migration of hazardous substances and to prevent human contact.

After abating these immediate threats, EPA may undertake remedial actions to clean up the more complex environmental problems at sites.

An example of a Federal emergency response to a release of a hazardous substance is provided.

Anatomy of a Hazardous Substance Response

Emergency: At a chemical manufacturing facility, chlorine is used to bleach dye from wastewater before discharging the water into a sewer. The chlorine is stored as a liquid in a pressurized, 30-ton capacity above-ground storage tank. At 4:30 a.m., Saturday, a control valve on the tank fails and liquid chlorine is released at a rate of about 200 pounds per hour (the reportable quantity for chlorine is 10 pounds). The pool of chlorine quickly vaporizes, resulting in a toxic cloud of chlorine gas drifting toward nearby residents.

Notification: The night supervisor discovers the release at 4:40 a.m. and dials 911 immediately to summon the local police and fire department. The supervisor then calls the National Response Center, State emergency response commission, and local emergency planning committee to report the release, as specified in the facility's response plan. Within seconds, the National Response Center relays the information to the predesignated EPA On-Scene Coordinator (OSC).

At the Scene

Response: The police and fire department arrive at the facility at 4:48 a.m. After surveying the scene, the fire department calls in its hazardous materials team for assistance. The EPA OSC—enroute to the site—telephones the fire department for details on the spill; about 100 pounds of chlorine have already been released and several local residents are indicating respiratory problems. The local responders begin a dual response: repairing the valve and evacuating the area. The OSC arrives at the site at 6:30 a.m. The local responders are unable to repair the valve and request Federal assistance. After conferring with the police and finding that evacuation is proceeding slowly, the OSC determines that the release is beyond the capacity and resources of the local responders, and that Federal assistance is needed. At 6:45 a.m., the OSC telephones the EPA Regional office to get backup response personnel and equipment. The OSC then calls the Environmental Response Team for special engineering expertise on sealing the valve and coordinates with State authorities to get National Guardsmen to assist with the evacuation. Chlorine response experts start to arrive on-site at 7:45 a.m.; they successfully seal the valve by 7:59 a.m. At 8:20 a.m., the Federal responders, wearing protective equipment, contain the remaining liquid chlorine to prevent further evaporation. More than 800 pounds of chlorine were released. The OSC arranges for trained medical personnel to be brought in to treat the residents for chlorine exposure. Once the area is secure, the OSC oversees site cleanup by contractors hired by the facility owners. . . .

> ***"[Environmental protection] cannot be accomplished without assistance and cooperation from the public."***

EPA plays a vital role in protecting our environment, but this cannot be accomplished without assistance and cooperation from the public. Some important responsibilities of local communities are to notify the National Response Center ((800) 424-8802) of any emergency incidents, be prepared for releases that do occur, assist with the response, and generally stay informed.

The Agency works with the public by informing local communities of emergency incidents and resulting response actions.

EPA Programs Reduce Air Pollution

by Eileen Claussen

About the author: *Eileen Claussen is the former director of the EPA's Office of Atmospheric and Indoor Air Programs. Claussen is now a special assistant to President Bill Clinton and senior director for global environmental affairs at the National Security Council in Washington, D.C.*

It is a pleasure to present a number of voluntary Environmental Protection Agency (EPA) initiatives relating to climate change that would reduce greenhouse gas emissions and air pollutants associated with the production of electricity by promoting private investment in energy efficiency. Greenhouse gases and air pollutants that would be affected include carbon dioxide (CO_2), methane (CH_4), nitrous oxide (N_2O), nitrogen oxide (NO_x) and sulfur dioxide (SO_2).

Energy-Saving Strategies

The EPA voluntary programs are based on the premise that measures to encourage full and more rapid information dissemination and adoption of improved technologies by market leaders can increase the productive use of energy and lead to increased reductions in greenhouse gas emissions. These measures can help in the following ways:

• an improved understanding of purchase price and total life cycle cost in decisions for selecting products;

• increased penetration of improved technologies in U.S. and world markets leading to earlier price reductions for efficient products;

• generally increased availability of cost-effective technologies, for a range of applications, for preventing, controlling, and mitigating greenhouse gas emissions and, wherever possible, for also recovering energy;

• earlier penetration by existing efficient products sending the signal to manufacturers that there is a market for such products, and eventually for even more

From Eileen Claussen's statement to the House of Representatives Committee on Energy and Commerce, Subcommittee on Energy and Power, March 3, 1992.

advanced technologies; and finally,

• improved information and accelerating demand for energy efficiency investments will encourage changes in utility regulations relating to potential efficiency investments.

We believe that these measures can have substantial effects. To address energy efficiency objectives, EPA has developed a five-part strategy: (l) corporate and government purchasing; (2) enhanced product markets; (3) regulatory and legal reforms; (4) development and demonstration of prevention, control, and mitigation technologies which, wherever possible, also recover energy; and (5) expanded international markets. EPA believes that this strategy of efficiently using energy resources can cost-effectively achieve reductions in emissions of greenhouse gases below the levels which they would otherwise reach in future years, while at the same time promote economic growth.

In the last two decades, U.S. energy consumption per unit of GNP [gross national product] has fallen by 30 percent, mostly as a result of the normal functioning of the market. "Green Lights" and related EPA programs targeted at electricity consumption are part of an overall movement to improve the efficiency of energy use, as well as utilities' integrated resource planning and demand-side management programs. We are coordinating closely with the Department of Energy on all of these efforts. The best way to explain how this strategy would work is to place it in the context of programs that EPA has already developed, or is currently developing, in each of these strategic areas.

Green Lights

Our flagship program in the area of corporate and government purchasing is Green Lights, which was formally launched on January 16, 1991. Green Lights is a voluntary pollution prevention partnership between EPA and corporations, States or local governments. Green Lights partners commit to evaluating current lighting needs throughout their facilities and installing energy-efficient lighting wherever it is profitable (as measured by the prime rate plus 6 percent). EPA provides technical assistance through building survey software, support to a product testing laboratory, information about manufacturers and financial assistance sources, and public recognition.

> ***"Efficiently using energy resources can cost-effectively achieve reductions in emissions of greenhouse gases."***

EPA believes this program will be successful, since over 400 participants have already signed up—nearly two billion square feet committed—which is more than the office space in New York, Los Angeles, Chicago, Houston, Dallas and Detroit combined.

While corporations have five years to complete their upgrades, results from preliminary installations look quite promising: reports received indicate that savings of 40 to 75 percent in lighting electricity consumption have been realized, exceeding some previously published technical estimates. Preliminary analyses indicate that investment in energy-efficient lighting resulting from utility including integral resource planning and demand-side management, as well as Green Lights, could lower greenhouse gas emissions from electricity production by millions of tons.

"Energy-efficient lighting . . . could lower greenhouse gas emissions . . . by millions of tons."

Why have these lighting upgrades not happened earlier, given that the investments in efficient lighting products are all profitable? There are a variety of reasons. The facilities manager is responsible for lighting—and usually also charged with making the elevators run and ensuring that the building temperature is comfortable. Even when this manager is educated that opportunities exist for improving efficiency in lighting and saving money, and able to sort through the various products and their claims, adequate and reliable information and the organizational support may not always be available to make the necessary investments.

Incentives for Efficiency

EPA's Green Lights program requires participants to establish efficient lighting as a strategic corporate decision undertaken at a high level within the corporation and it also provides information regarding the coordination of options available from the existing range of lighting products as a system. The software, product testing and financing information support provided by EPA helps the corporation make the most attractive upgrades. And the public recognition afforded by the Green Lights program, as well as the significant potential savings, provides incentives for corporate leaders to commit to improving their energy efficiency. At the same time, thanks to changes in State regulatory climates, utilities are starting to make major investments in energy efficiency measures in conjunction with their customers, increasing the attractiveness of these programs.

EPA is currently developing several other programs that could operate similarly to the Green Lights model with regard to corporate and government purchasing; green commercial buildings; and green energy corporations.

The Green Buildings program would focus primarily on building-shell improvements. As with lighting, efficient technologies exist today that could significantly reduce building energy needs. These technologies include variable speed drives for air handlers and chillers, reflective painting on rooftops and window improvements.

Following up on interest from some of our Green Lights partners, EPA is also exploring the possibility of a holistic program to stimulate corporate energy efficiency across the board—in other words, a green energy corporation program that would require participants to ensure that all equipment purchases are consistent with the principles of life-cycle profitability.

EPA's Green Lights—and similar "green" programs—demonstrate that significant pollution prevention may be achieved by providing information and technical advice to corporations and government offices.

Enhanced Product Markets

A second aspect of EPA's approach is to provide signals to manufacturers that markets exist for the most efficient technologies that manufacturers can produce. EPA's premiere example of enhanced product markets is the "Golden Carrot" super-efficient refrigerator program. The Golden Carrot program takes the rebates utilities have agreed to provide to their customers who buy efficient refrigerators after 1994 and aggregates them into a single bid pool. Manufacturers then bid on this pool of rebate money. The bid pool is offered to the manufacturer that is able to provide the greatest number of (non-chlorofluorocarbon using) super-efficient refrigerators at the least cost by 1994 (or 1995).

Under present law, the Department of Energy (DOE) standard is applied to all refrigerators as a minimum standard. The winning refrigerator model is required to be 25 percent more efficient than the 1993 DOE appliance standard, and may be as much as 50 percent more efficient. . . . The first super-efficient refrigerators will be shipped as early as 1994. The rebates will be awarded directly to the manufacturers.

Opportunities for many other Golden Carrot programs exist. In fact, an organization has recently been formed by EPA, utilities and conservation groups—the Consortium for Energy Efficiency—to coordinate Golden Carrot and other utility demand-side management programs. The consortium strives for similar approaches among utility programs and will send clear and consistent signals to equipment manufacturers, distributors and retailers. In the future, the Consortium for Energy Efficiency will facilitate information exchange and coordinated purchasing of super-efficient equipment.

EPA has identified a large number of opportunities for future Golden Carrots, including heat pumps, clothes washers, clothes dryers and solar water heaters.

> ***"Markets exist for the most efficient technologies that manufacturers can produce."***

Another area in which EPA programs are enhancing markets for efficient products is office equipment. Computer equipment is the fastest growing electricity load in the fastest growing electricity-using sector: commercial. EPA's Green Computers program is a voluntary effort just underway with manufacturers of computer equipment to produce and market energy-efficient desktop computers. EPA will provide the manufacturers with a label indicating that this equipment uses less electricity. Manufacturers can use this label to indicate to consumers that their computer is energy-efficient. We believe that this concept can also be extended to office copiers, fax machines, and other energy-using consumer products. By amplifying market signals for efficient products, these programs can stimulate real savings both in terms of energy and reduced greenhouse gas emissions.

Regulatory and Legal Reforms

Step three of EPA's strategy is to help ensure that corporations are not penalized for efficient use of energy. The programs I have just outlined will not function effectively if utilities lose money as their customers become more energy-efficient. As long as utilities have a regulatory structure which provides incentives to *sell* kilowatt hours and disincentives to *save* kilowatt hours, national efforts toward energy efficiency will be at odds with the shareholder objectives currently in place in most utility jurisdictions. For federal policies to be effective, they must not be inconsistent with the objectives of utilities around the country.

Since the mid-1980's, the Department of Energy has had a program of working with the National Association of Regulatory Utility Commissioners and individual utilities in support of Integrated Resource Planning, an approach that seeks to ensure that both supply and demand options are evaluated and rewarded similarly. This effort, together with pioneering work in Massachusetts and California by utilities and State commissions, has led to a sea change in State regulatory treatment of energy efficiency investments by utilities.

> ***"Another area in which EPA programs are enhancing markets for efficient products is office equipment."***

The EPA has engaged in a widespread outreach effort that complements what is already underway. EPA is meeting with State utility commissions and governments to discuss the environmental and economic benefits of regulatory reforms. Relevant issues include de-coupling profits from sales and "shared savings" plans, in which utilities have positive economic incentives to aggressively pursue efficiency improvements. Conservation is also emphasized as a strategy for complying with the acid rain provisions of the Clean Air Act amendments. EPA hopes to continue this positive working relationship with members of the State regulatory community.

Reducing Methane Emissions

While not based on products or corporate purchasing per se, profitable options identified by EPA for reducing emissions of methane to the atmosphere also fit into the strategy. Methane is an attractive focus for reducing greenhouse gas emissions because it is more effective at trapping heat than carbon dioxide

(CO_2) and because it has energy value—so profitable systems can be designed to recover or better utilize this methane. This is possible with methane from landfills, methane from the management of animals (dairy and swine farming) and methane from coal mining. EPA is working to enhance market opportunities for methane recovery in these areas. This includes demonstrating profitable "on-site" energy generation for farms (while solving non-point source runoff problems) and working to remove barriers to energy recovery at landfills and coal mines. For example, property rights issues currently limit the recovery of "pipeline quality gas" emitted during coal mining in Appalachia. The underlying issues need to be examined with a view toward finding a solution.

EPA is involved in three commercial-scale demonstrations to control methane and recover energy. Two of these are the world's first applications of a fuel cell for this purpose on landfill gas and also on anaerobic digester gas. A third demonstration is directed toward advanced pregasification for deep coal mines.

Expanded International Markets

Finally, EPA is helping develop international markets for efficient American technologies. At present, we have projects underway in China (refrigerators and coalbed methane), Russia (natural gas pipelines) and Poland (coalbed methane). These technology cooperation projects will contribute significantly to global environmental improvement, and potentially create large markets for domestic goods and services and will thereby decrease prices for these goods and services, both abroad and within our own borders.

Voluntary programs—following the model of EPA's Green Lights program—have demonstrated that they can reduce greenhouse gas emissions in a cost-effective way. We believe that by improving information dissemination about efficiency and continuing technology development and demonstration, cost-effective reductions in emissions of greenhouse gases can be achieved while promoting economic growth.

The EPA Helps Prevent Toxic Pollution

by Linda J. Fisher

About the author: *Linda J. Fisher is the former assistant administrator of the EPA's Office of Prevention, Pesticides, and Toxic Substances. Fisher is now an attorney at the law firm of Latham and Watkins in Washington, D.C.*

In the last several years, substantial new responsibilities have been added to [the EPA's] traditional toxics program functions. The largest growth area has been the collection, analysis and dissemination of data for the Toxics Release Inventory (TRI) under the Emergency Planning and Community Right-to-Know Act (EPCRA) of 1986. This is proving to be one of the most significant tools that the Agency and State and local governments have for identifying potential problems and remedies for toxic chemical releases. Over 80,000 facility reports are processed each year and each of these reports requires expert scrutiny to insure data quality. The Pollution Prevention Act of 1990 greatly expands the coverage of the information collected for the TRI, which substantially increases the complexity and cost of maintaining this system.

Strategies Against Lead

This office also has principal responsibility for implementation of the Agency's lead strategy. This includes a wide variety of activities to help identify and respond to geographic hot spots of lead exposure to children; promote public outreach and education programs to reduce human exposure; and develop standards, guidelines and technical assistance for management and abatement of in-place lead. These activities are being accomplished through inter-Agency efforts involving the Department of Housing and Urban Development (HUD), the Department of Health and Human Services (HHS), and State agencies, as well as other EPA offices and our regions. Increased concern over the health effects of lead exposure to children have created a demand for Agency

From Linda J. Fisher's statement to the Senate Subcommittee on Toxic Substances Research and Development, March 25, 1992.

information and action to which we must respond.

The new 33/50 Project has become our showcase program for promoting non-regulatory approaches to toxic emission source reductions. It encourages industry to reduce voluntarily the emissions of 17 highly toxic chemicals by 33 percent by the end of 1992 and by 50 percent—or 700 million pounds—by the end of 1995. As of January 1992, 734 companies agreed to participate in the program, and they made commitments to reduce their toxic emissions collectively by 304 million pounds. While it is too early to claim victory, every indication so far suggests that this will be one of the most successful voluntary programs the Agency has ever sponsored, and it may represent a whole new way of doing business with industry.

"The new 33/50 Project has become our showcase program for promoting . . . toxic emission source reductions."

Finally, our role as the Agency advocate for pollution prevention is our most recent new assignment. It has led us to change the name of the Office of Toxic Substances to the Office of Pollution Prevention and Toxics (OPPT). This responsibility for implementation of the Pollution Prevention Act and for fostering pollution prevention, whenever cost effective, builds on our long-standing capacity for cross-media risk evaluation and management.

As our responsibilities have grown to include these new higher risk and prevention-oriented program areas, we have shifted some of our resources from the more traditional TSCA [Toxic Substances Control Act] programs to accommodate them.

Although our range of activities has broadened, we are still intent on getting the most out of the tools available to us through TSCA. In 1990, for example, OPPT launched a major revitalization of its existing chemicals program to increase the number and effectiveness of actions that we take to reduce the potential health and environmental risks posed by chemicals currently in production. . . .

Guiding Principles for the Future

As a result of our changing program and in the aftermath of the asbestos court decision [in 1991 faulting EPA techniques in assessing asbestos exposure], we have examined what the basic objectives of a toxics program should be. While we want to utilize the traditional regulatory tools which TSCA provides, we also see opportunities to use our statutory authorities in fresh ways, consistent with new trends in environmental management. OPPT has identified the four key principles below to help guide future efforts to reduce health and environmental risks from toxic substances under the various statutes that we administer.

1. *Promoting Pollution Prevention.* Under our new mandate, a primary mission of OPPT is to encourage the use of pollution prevention principles. There are several models we can use to accomplish this, ranging from encouraging

voluntary initiatives to incorporating cost effective prevention alternatives into regulations. OPPT will be working with other EPA offices to identify regulations that should be targeted for pollution prevention options at the same time that it looks for additional ways to develop creative non-regulatory strategies for implementing public health and environmental goals.

We will be carrying the pollution prevention message outside the Agency as well, through provision of information, training and education, and the extension of grant assistance. For example, in 1991 we announced a grant to the University of Michigan for establishment of a National Pollution Prevention Center. This center's primary purpose is to develop model undergraduate and graduate curricula for engineering, natural resources and business schools. The ultimate aim is to educate those who will design, develop, and market chemical products about incorporating the prevention ethic into their work. We are also trying to further stimulate existing incentives that manufacturers have to design chemicals using cost effective alternative synthetic pathways for manufacturing chemicals with an eye on the environmental implications of their choices.

EPA's traditional focus has been the industrial sector of our economy. In the future, our pollution prevention initiatives will carry us into new areas as we develop pollution prevention strategies for sectors such as agriculture, energy and transportation. We will be working with other Federal agencies and with outside groups to promote the idea that pollution prevention can be an environmentally sound and economically prudent way to approach decisions in all sectors of our society.

"The adoption of a pollution prevention ethic is a logical development."

Safer Chemicals

2. *Using TSCA to Promote the Design, Development and Application of Safer Chemicals.* For OPPT, the adoption of a pollution prevention ethic is a logical development, given the focus of our program on improving environmental protection through changes in the manufacture, processing, and use of chemicals in our society. Fundamentally, our role is to push for use of safer chemicals and processes in the basic operations of the industrial sector.

Our new chemicals program illustrates the essence of prevention. In addition to keeping chemicals that will pose significant risks out of the marketplace, it also serves as a mechanism for identifying promising alternatives to existing chemicals. We have used it in this way by revising our PMN [premanufacture notification] form to allow submitters to show how a proposed new chemical might serve as a better substitute for an existing chemical.

In the existing chemicals program, we see an increasing emphasis on evaluating clusters of chemicals that are used in particular industrial processes with an eye toward promoting safer chemicals and technologies at a reasonable cost. We believe this approach will eventually replace the tendency to focus on single chemicals, which may sometimes lead to inadequate consideration of the risks of substitutes.

> ***"OPPT . . . is now coordinating a multi-media, Agency-wide strategy to address health risks from lead."***

3. *Providing Stewardship for High Risk Chemicals.* Although pollution prevention will be a guiding principle, we still face the task of managing several high risk chemicals, such as lead, asbestos and PCBs, that have been widely used for years in buildings and equipment.

Asbestos and PCBs are sometimes referred to as OPPT's "first generation" programs. These substances have been around a long time, their health risks are well documented and widely known, and the management of problems caused by past uses will be required well into the future.

We have established, for example, a comprehensive framework for addressing asbestos in our schools, including a major regulatory program (under Title II of TSCA) and a national accreditation network for inspection and abatement personnel administered by the States. Our recent guidance on managing asbestos in place identifies proper practices and procedures for public and commercial, as well as school, building managers to reduce exposure in their facilities.

While lead is ubiquitous in the environment and its risks are also well documented, some of our initiatives to deal with problems related to past usage are new or still under development. OPPT has developed and is now coordinating a multi-media, Agency-wide strategy to address health risks from lead. We have in the past given detailed testimony on the lead strategy before this Subcommittee. Activities in the lead strategy range from new materials to help parents reduce lead exposure to children in their homes to studies of the efficacy of various lead abatement alternatives and new TSCA regulatory investigations on lead in plumbing fittings, fixtures, and solder.

Information and the Right to Know

4. *Acquiring and Disseminating Information About Chemical Risks and Pollution Prevention.* TRI, established in Section 313 of EPCRA, has stimulated pollution prevention efforts. It provided an opportunity for both citizens and industry to understand the volume of toxic chemical emissions, and the result has been that both groups are working to reduce them.

We have recognized that the right-to-know objectives of TRI can be merged with the information-gathering powers of TSCA to make additional data available to all those who have a role in environmental decision-making. . . .

In addition to the information gathering and dissemination efforts under TRI and TSCA, we also established a pollution prevention clearinghouse in 1990 that contains technical, policy, legislative, and financial information about source reduction and recycling efforts in the United States and foreign countries. Clearinghouse information is free and easily accessible by phone, computer, or mail. To help get the information to those who can use it most, we have been actively encouraging State governments and industry to use the clearinghouse.

We see the acquisition and dissemination of information to other levels and agencies of government, to industry and to the public as being one of our prime responsibilities. We are committed to using TRI as a model for carrying out this obligation. As we move forward with a revitalized chemicals program, we will embrace principles of pollution prevention, safer chemicals, cost effectiveness and information-sharing to guide our efforts.

Defining Success

We define success in our toxics program as preventing or reducing risk, and noticeable changes in industry and public behavior is our measure. I believe we are seeing those changes in a variety of new areas: in TRI emission reductions, in commitments to the 33/50 Program, in Section 8(e) Compliance Audit Program participation.

Success, in many of these new instances, is not always tied to TSCA regulations. While we will not step back from our statutory responsibilities under TSCA, we cannot measure success simply by the size of our regulatory agenda. If 33/50 Program commitments, for example, are realized, this will likely achieve more environmental results than many of the [TSCA] Section 6 rulemakings that we have considered or implemented in the past. . . .

Effecting behavioral change is never easy, and it will not be easy in the area of toxic substances. However, I believe that a balanced combination of voluntary approaches and traditional and innovative regulatory programs, which we are now pursuing, is most appropriate.

The EPA Fails to Protect the Environment

by William Sanjour

About the author: *William Sanjour, a policy analyst for the EPA, has worked for the agency for more than twenty years.*

For the past 20 years I have worked for the Environmental Protection Agency. There I have had to choose between being a "good soldier" and obeying orders or being a "good citizen" and obeying the law. I have not, I'm afraid, been a very good soldier. When I came to the then-new agency, I hoped to do something useful and constructive. In 1974 I was made a branch chief of the Hazardous Waste Management Division. The studies I supervised there played an important part in the passage of the Resource Conservation and Recovery Act (RCRA) of 1976, the first federal law regulating toxic waste. I was also in charge of drafting rules for the treatment, storage, and disposal of hazardous materials. In its preoccupation with inflation, however, the Carter administration in 1978 took steps to protect industry by removing the teeth from those regulations. At first I fought from the inside to preserve the true spirit of the legislation. As a result, in 1979 I was transferred to another position, with no duties and no staff. I became an outspoken EPA critic—a whistleblower—and have been one ever since.

Ordinary Citizens

In that role I spend much of my spare time meeting with grassroots environmental groups. Their members frequently ask me why the Environmental Protection Agency does not seem particularly interested in protecting the environment. The question usually comes from people who are dealing directly with the EPA for the first time, ordinary citizens with ordinary political views and lifestyles who suddenly find themselves living close to a hazardous-waste facility, incinerator, or nuclear-waste dump. These are people who started out with a

William Sanjour, "In Name Only." Reprinted with permission from the September/October 1992 issue of *Sierra* magazine. Sanjour's longer, more detailed critique of the EPA is available for $15 from the Environmental Research Foundation, PO Box 73700, Washington, DC 20056-3700.

strong faith in their country and its institutions, who had always thought of the EPA as the guys in white hats who put the bad polluters in jail. "If there were anything wrong with it," they say, "the government wouldn't let them do it."

To their surprise, these folks find that the EPA officials, rather than being their allies, are at best indifferent and often antagonistic. They find that the EPA views *them*, and not the polluters, as the enemy. Citizens who thought that the resources of the government would be at their disposal find instead that they have to hire their own experts to gather data on the health and environmental impacts of proposed facilities, while the government sits on the same information—collected at public expense. And if these folks want to go to court, they have to run bake sales to hire attorneys to go up against government lawyers whose salaries are paid by the taxpayer.

To understand why the Environmental Protection Agency is the way it is, you have to start at the top, and since the EPA is part of the executive branch, that means the White House. The president (any president, regardless of party) and his immediate staff have an agenda of about a half-dozen issues with which they are most concerned. These are usually national security, foreign affairs, the economy, the budget, and maybe one or two others: Call them Class-A priorities. All others—housing, education, transportation, the environment—are in Class B.

Presidential Priorities

The president expects performance in Class A. He will expect the military to be able to deploy forces anywhere in the world when an emergency arises—and if it isn't, he will bang heads until it is. If Congress doesn't support his budget, he will call the budget director into his office and pound his fist on the table. But can you picture the president bringing the secretary of transportation into the Oval Office and yelling because of poor bus service in Sheboygan? Or summoning the administrator of the Environmental Protection Agency in and chewing him out for pollution in the Cuyahoga River? I can't. The president expects performance in Class A; in Class B he expects only peace and quiet.

But regulatory agencies, by their very nature, can do little that doesn't adversely affect business, especially big and influential business, and this disturbs the president's repose. The EPA, for instance, cannot write regulations governing the petroleum industry without the oil companies going to the White House screaming "energy crisis!" If it tries to control dioxin emissions, the *New York Times* (whose paper mill in Canada has been sued for dumping dioxin into the Kapuskasing, Mattagami, and Moose rivers) writes nasty editorials. If it tried to enforce the Clean Air Act, polluters ran to Vice-President Quayle's Council on Competitiveness for "regulatory relief." Agency

"[Citizens] find that the EPA views* them, *and not the polluters, as the enemy."

employees soon learn that drafting and implementing rules for environmental protection means making enemies of powerful and influential people. They learn to be "team players," an ethic that permeates the entire agency without ever being transmitted through written or even oral instructions. People who like to get things done, who need to see concrete results for their efforts, don't last long. They don't necessarily get fired, but they don't advance either; their responsibilities are transferred to others, and they often leave the agency in disgust. The people who get ahead are those clever ones with a talent for procrastination, obfuscation, and coming up with superficially plausible reasons for accomplishing nothing.

Sludge Warnings Ignored

For example, the EPA used to grant billions of dollars for the construction of local sewage-treatment plants. These plants generated a sludge that the EPA recommended for use as a fertilizer. In 1974 I pointed out to my managers that there was considerable evidence from Department of Agriculture studies that some municipal sewer sludge contained poisons that could be transmitted to people when it was used as fertilizer. I proposed regulations to control the problem. This notion was very unpopular with the burgeoning sewage-plant-construction industry and its promoters within the EPA. The responsibility for this issue was taken out of my hands and transferred to a committee, which studied sludge regulation for a year and did nothing other than recommend further study. For this they all received medals and cash bonuses as "outstanding performers."

"People who like to get things done [at EPA] . . . don't last long."

In the past 18 years this story has been repeated many times. Hundreds of people in the EPA have advanced their careers—and spent tens of millions of dollars—by busily drawing up work plans, attending meetings, making proposals, writing reports, giving briefings, conducting studies, and accomplishing nothing. Today the problem of how to regulate sewage sludge has still not been resolved.

At this point you may protest that the EPA *has* written many regulations, that it has in fact reduced pollution in many areas, cleaned up Superfund sites, and collected millions of dollars in fines from polluters, some of whom have even been sent to jail. Yet in most cases the agency had to be coerced into meaningful action. More often than not, the EPA actually *opposes* the passage of tough environmental laws, and organizations like the Sierra Club have to sue in federal court just to make the agency do what it is funded for and is legally required to accomplish. For example, when I was writing guidelines for the government's procurement of recycled materials, I was told that a proposed regulation would not even be considered for the administrator's signature unless there was a court-ordered deadline. With my encouragement, several organizations sued the EPA in order to get the regulations out.

On another occasion I was in charge of writing regulations for the management of hazardous-waste landfills, which RCRA required to be issued in 1977. When Gary Dietrich, my boss, gave orders delaying the process, I warned him that we would miss the legal deadline. He laughed. "Nobody ever got thrown into jail for missing a deadline," he said.

"Nobody was sent to jail for defying the court's orders or for not implementing the law."

He was right. I was taken off the job. Again I contacted an environmental organization, which sued the EPA. The court imposed another deadline. The EPA missed that one, and the judge set another. They missed that one too, and many, many more, but nobody was sent to jail for defying the court's orders or for not implementing the law. On the contrary, many were well rewarded. Meanwhile, the public was exposed to poisons leaking out of countless unregulated hazardous-waste dumps. Dietrich later left the agency to be a consultant for Waste Management, Inc.

(This leads us to what I call Dietrich's Law: "No one in the EPA is ever sent to jail, or loses his job, or suffers any career setback for failing to do what the law requires." And the corollary: "Many people ruin their careers in the EPA by trying to do what the law demands.")

The landfill regulations were finally issued in 1982, five years after they were due. They were riddled with loopholes, such as the final say given to politically appointed regional administrators in setting safety levels of toxic materials. Even so, the press hailed the EPA's heroic achievement—although the bloom quickly faded. After hearing testimony from me and many others, Congress was convinced that the regulations were too weak, and passed a new law in 1984 requiring tougher standards. This time it added a "hammer" provision: If the EPA missed the deadline, then all the wastes from a long list of chemicals would be banned from landfills. For the first time I can recall for regulations of this magnitude, the EPA met its deadlines. Why? Because in this case, hazardous-waste firms would have been hurt if the rules weren't issued. The EPA is simply more concerned with protecting the industries it is supposed to regulate than in protecting the public interest.

The Wrong Goals

Does this mean that the EPA has cynically abandoned the environment for the sake of the powerful hazardous-waste lobby? Actually, most people in the agency sincerely equate the waste-management industry with protection of the environment, and see the industry's opponents as anti-environmental NIMBYS [not-in-my-back-yard]. They forget, however, that commercial hazardous-waste management is primarily a business, and as such it aims to maximize income and minimize costs. Income is produced by taking in wastes through the gate; waste is money, and the more the better. Costs are incurred by treating the waste so that it won't poison people and the environment. Obviously, these goals are diametrically opposed to what should be those of the EPA: to reduce the production of hazardous wastes and to maximize protection of human health and the environment.

"Local citizens found that the permit originally issued by the EPA was full of... outright violations of the law."

The EPA's confusion on this matter is well illustrated in the case of the world's largest hazardous-waste incinerator, now under construction in East Liverpool, Ohio, an already heavily polluted area surrounded by homes and schools and subject to frequent thermal inversions. (Behind the project is a consortium of investors put together by Arkansas billionaire Jackson Stephens, a golfing partner of Dan Quayle and contributor of hundreds of thousands of dollars to George Bush's presidential campaigns.) Local citizens found that the permit originally issued by the EPA was full of irregularities and outright violations of the law. Thus, when the incinerator operator asked for a permit modification to install a spray dryer (a device many technical experts felt was unsafe), the permit would ordinarily have had to be issued again, not just modified.

However, given the public mood, this was likely to result in long delays, if not revocation of the permit. The incinerator operator told the Ohio EPA that he couldn't "risk any appeals." The Ohio EPA agreed, saying that "if there is a way to authorize this change without a formal permit change, we should try to do so." William Muno, acting director of the EPA's waste-management division in Chicago, followed his instincts: At a meeting with congressional staffers on November 12, 1991, he said that he would not order a permit change because the EPA "had to treat our constituents [i.e., the incinerator operators] in a fair and equitable manner."

(Influence peddling in this arena does not stop with the EPA. The executives and lobbyists of the waste-management companies are in constant touch with the White House, members of Congress, state legislators, state environmental-protection agencies, the press, and national environmental organizations. The Audubon Society, the National Wildlife Federation, and the Conservation Foundation all have top executives of the waste-management industry on their boards.)

Growth Industry of Today

Waste management is the growth industry of the late 20th century. It has become very rich through its ability to control the regulators who are supposed to control it—and it shares this wealth with its benefactors. Government bureaucrats soon learn that, while crossing the industry can get them into a lot of trouble, cooperating with it has many rewards—high among these the hope of lucrative future employment. Indeed, rather than the environmental enthusiasts who flocked to the EPA in its early years, the agency is now full of careerists who view their jobs as stepping-stones to bigger and better things. Scores of federal and state employees have gone on to careers in the hazardous-waste industry, including three out of the five EPA administrators. (Of the other two, one left the agency in disgrace and one was a millionaire already.)

No one is more closely associated with the revolving door at the EPA than William Ruckelshaus, appointed the agency's first administrator when it was created in 1970. When he left the EPA in 1973, Ruckelshaus became senior vice-president and director of Weyerhaeuser, the huge timber and paper company and target of many environmental groups. He served as EPA administrator a second time from 1983 to 1985. Between and after his two terms he was a director of several companies concerned with EPA regulations, including Monsanto, Cummins Engine Company (a diesel-engine manufacturer), Pacific Gas Transmission, and the American Paper Institute.

After his second stint at the agency, he formed a consulting firm called William D. Ruckelshaus Associates, which was then hired by the Coalition on Superfund, an organization seeking to weaken the Superfund law by absolving polluters of strict legal liability for their actions. The coalition included such Superfund polluters and their insurers as Monsanto, Occidental Petroleum, Alcoa, Dow Chemical, AT&T, Du Pont, Union Carbide, Aetna Insurance, and Travelers Insurance. Assisting Ruckelshaus were Lee Thomas, his hand-picked successor as EPA administrator, and William Reilly, then head of the Conservation Foundation. (Ruckelshaus and Thomas helped fund Reilly's organization to produce studies in support of the coalition's position.)

"The agency is now full of careerists who view their jobs as stepping-stones to bigger and better things."

The Revolving Door

Ruckelshaus went on to become CEO of Browning-Ferris, the second-largest waste-management company in the United States, for a guaranteed minimum annual salary of $1 million. Browning-Ferris had a dreadful environmental record and had been hit with millions of dollars in fines. Ruckelshaus was supposed to clean up the company's reputation, but the appointment did more to tarnish his.

When George Bush ran for president in 1988, Ruckelshaus was his environmental advisor, and was able to install his protégé Reilly as EPA administrator, and former Ruckelshaus Associates vice-president Henry Habicht as deputy administrator. Thus the two top executives of the EPA were placed by the head of a company that is a major polluter, heavily regulated by the agency and a principal party responsible for many Superfund sites.

People outside the agency often assume that the national environmental groups have a stronger influence within the EPA than does industry. The revolving door explains why this is not the case: Industry can offer EPA employees things that environmentalists cannot, especially high-paying jobs. It also offers generous contributions, over or under the table, to almost anybody who will take its money.

(Waste Management, Inc., has one of the largest political-action committees in the country; between 1987 and 1988 it contributed $430,000 to congressional candidates. Other examples of WMI's largess include flying a politician in a corporate jet to visit a WMI facility and giving him a cash gift of $10,000; giving a congressional staffer a $2,000 "honorarium" to visit a WMI facility; and paying an outright $3,000 bribe to a local commissioner in Florida.)

Environmental groups tend to regard the EPA as an institution, dealing with it through congressional committees, the courts, and top agency executives. Industry does the same, of course, but it also interacts with individual EPA employees at every level, working directly with the field inspectors and permit writers responsible for making particular decisions. When I was in charge of writing regulations I was the object of this courtship, showered with flattery, meals, trips, and hints of future employment. People who cooperate with industry also find that its lobbyists will work for their advancement with upper management. Those who don't cooperate find the lobbyists lobbying for their heads.

How to Reform EPA

What can be done, then, to make the EPA serve the public interest? Appointing an energetic administrator and giving him or her a lot of money and authority will not work. If, as is usually the case, the president demands only peace and quiet, more funding and power will only make that easier to deliver. The head of the EPA will not be effective unless the president wants effectiveness. In my 20 years with the agency, that happy situation existed only under Richard Nixon during William Ruckelshaus' first term as administrator. Presidents Ford and Carter were too concerned with the economy, and paid only lip service to EPA regulations. Ronald Reagan didn't even bother with lip service, appointing hooligans to run the agency. When Anne Gorsuch was finally forced to resign in March 1983 over her political manipulation of the Superfund program, Ruckelshaus was brought back as Mr. Clean; but his second administration was a flop. The man who had been so effective under Nixon was a dud under Reagan, which shows that the president is key to making the

> ***"Those who don't cooperate find the lobbyists lobbying for their heads."***

EPA work. Under George Bush, we reverted to the days of lip service, with no real support. This is, I'm afraid, what scientists would call the EPA's "equilibrium state."

To achieve genuine reform, realism must replace idealism. We have to deal with what the EPA really is and what we know about it, rather than what we would like it to be. This will require narrowing the agency's discretionary power and transferring as much authority as possible into the hands of the public. Following are some suggestions to those ends:

"The EPA works more diligently to protect the industries it regulates than to protect the public."

Hammer Provisions. As we saw above, the secret of these blunt legislative instruments is their tacit recognition that the EPA works more diligently to protect the industries it regulates than to protect the public. Hammer provisions can also be used to enforce goals, which at present are meaningless, since nothing happens if they aren't met. But suppose the EPA administrator and other top political appointees were hired with the proviso that unless specific targets were met—a 10-percent reduction in hazardous-waste generation within a year, say—they would lose their jobs. Perhaps then the administrator would spend more time in achieving goals than in making speeches about them.

Effective Provisions

Liability. Civil-liability provisions are another great, unbureaucratic instrument for reform. Such provisions in Superfund actually did more to change the way industry handles its hazardous waste than any other act of Congress—and this came about almost by accident. Congress, as is often the case, was vague and ambiguous when it defined the liability of polluters for the damage and cleanup costs of old dump sites. In interpreting that fuzzy language, however, the courts determined that liability is "strict." This means that no showing of fault is necessary to establish responsibility, little proof of the relationship between cause and effect is required, and each liable party is potentially responsible for the entire cleanup.

These provisions are so effective that industry and its insurance companies have spent millions trying to get rid of them. They would prefer that the funds for cleanup be pooled and paid out on a "no fault" basis. (There's nothing like strict liability to convert capitalists to socialism.) The fear of liability is a far greater incentive to industry to do the right thing than is fear of the EPA.

Regulations. The EPA is a wimpish regulator. Take the case of hazardous-waste incinerators. It makes no difference what you as a private citizen may see, feel, or smell coming out of the smokestack. Emissions could melt the paint off your house or force you to wear a gas mask, but that would have no bearing on the EPA's enforcement, which relies instead on data supplied by the incinerator operators. Ironically, agency officials often don't know how to interpret this data themselves, and have to depend on the company being investigated to do it for them.

Since little attention is paid to the content of the waste being burned, every now and then an incinerator explodes, as happened to a Chemical Waste Management incinerator in Chicago. In Kentucky, Don Harker, former head of the state's waste-management division, was fired because he tried to revoke the permit of the Liquid Waste Disposal (LWD) incinerator at Calvert City. "I don't know what LWD has burned," he said. "I don't think LWD knows what it has burned. I don't think anyone does."

Avoiding Action

Even if an inspector finds a violation, this only triggers a lengthy, complex process with many levels of warning, review, appeal, negotiation, and adjudication before any action is taken (or, more often, avoided). Compare this with what happens when you park under a "No Parking" sign. A policeman writes a ticket, and you can either pay the fine or tell it to the judge. If the EPA wrote the rules for parking violations, the policeman would first have to determine if there were sufficient legal parking available at a reasonable cost and at a reasonable distance, and would then have to stand by the car and wait until the owner showed up so that he could negotiate a settlement agreement. This is what comes of Congress giving the EPA administrator broad discretionary power to write and enforce rules. We would be better off if he were more like a cop.

> ***"We would be better off if [the EPA administrator] were more like a cop."***

Bad-Boy Laws. Several states have laws that bar them from doing business with chronic offenders. Unfortunately, these laws are usually discretionary and are rarely invoked in hazardous-waste cases. If they were, all of the big commercial hazardous-waste firms would be out of business in those states—which is why the laws are not used. I would like to see a mandatory federal bad-boy law applied to the licensing of hazardous-waste sites and to the awarding of Superfund contracts.

Of course, EPA officials always argue that if we close down the big commercial operators, there will be no one left to run the hazardous-waste business. That's like saying that we have to let racketeers run gambling casinos because no one else knows how. The hazardous-waste business is extremely profitable, and there are plenty of honest businesspeople who would love to get a foot in the door. There's no reason to tolerate crooks.

The Revolving Door. It should be perfectly clear that a person in a regulatory agency who views the agency as a stepping-stone to a better-paying job cannot serve the public faithfully. Yet Congress has never passed a law restricting persons in regulatory agencies from going to work for the companies their agency regulates. I would propose a law forbidding political appointees and senior government employees from accepting any form of direct or indirect compensation from any person regulated by their agency for a period of five years after they leave government service. The number of years could be reduced for lower-ranking civil servants. This law should, of course, include lawyers.

The common argument that this would keep good people from entering government is nonsense. Good people do not use government service as a means of getting rich quick. A revolving-door rule would keep out the ambitious careerists who now permeate the federal bureaucracy, and let in men and women with a real desire to serve their country.

Conflict of Interest. The revolving door is an individual conflict of interest, but there is a closely related institutional conflict in the EPA and some other regulatory agencies. The former Atomic Energy Commission, for example, had responsibility for regulating nuclear-power plants, but it also promoted the use of nuclear power. Similarly, the EPA has the responsibility for regulating hazardous-waste facilities, but also promotes the siting of these facilities. When the EPA assures the safety of a proposed hazardous-waste installation, citizens living nearby never know which hat it is wearing, regulator or promoter.

An easy solution is for Congress to pass a one-sentence law: "No regulatory agency may spend appropriated funds to promote the use of the products or services that it regulates."

Provide Incentives

The Carrot and the Stick. Every time the EPA falls short of expectations, it complains that it doesn't have enough money, people, authority, or time, so Congress and the courts give it more of everything. What kind of incentive is that? When I was a manager, if a subordinate failed to do an assigned task and gave me the excuse that he didn't have enough resources or time, I would ask him if he had requested the resources and time, and if he had warned me that the task could not be completed unless he got them. If the answer was no, he was in trouble. Congress should adopt the same standards for the EPA.

> ***"A revolving-door rule would keep out the ambitious careerists who now permeate the federal bureaucracy."***

One guaranteed quick fix would be to reward whistleblowers. At present, a person who calls attention to waste, fraud, or abuse at high levels of the EPA can only look forward to harassment and isolation, with no hope of promotion or even a responsible position. Congress ought not only to protect whistleblowers, but to reward them when their charges prove correct. This would greatly increase the number of whistleblowers and decrease the amount of waste, fraud, and abuse. If you think rewarding public servants for doing their duty is excessive, consider the cost of failing to do so.

Consent Agreements. One of the most egregious abuses of discretionary authority by the EPA is the use of consent agreements to settle regulatory violations. A consent agreement is like a plea bargain, a contract wherein a defendant agrees to stop an illegal activity without admitting guilt. Consent agreements by the EPA usually result in the defendant paying a fine and promising to sin no more.

"I believe the right approach is to reduce the power of institutions."

A big problem with consent agreements is that they are drawn up in secret, with no public review. While they usually concern cleanup of dumps or hazardous-waste spills, the injured community does not participate. This secrecy is an open invitation to corruption and abuse. Polluters with good connections and good lawyers are able to get consent agreements that grant them all sorts of privileges to which they are not entitled, in exchange for paltry fines. A good example is what happened when Chemical Waste Management was denied a permit to store carcinogenic polychlorinated biphenyls (PCBs) at its hazardous-waste dump at Emelle, Alabama. They stored them there anyway, and got caught. An eventual consent agreement fined ChemWaste far less than it made by its illegal action, and threw in a PCB-storage permit. The agreement also exempted the firm from punishment for any other past violations, even those that had not yet come to light. In short, for a $450,000 fine, ChemWaste received waivers worth more than $100 million. Congress should limit the scope of consent agreements, require that they be made public, and require the court to hear from any past or potential injured party before signing.

Power in the Right Hands

Institutions are made up of people. Behind the great and powerful Oz is a fragile little man pulling the levers. Because it must be implemented by weak and fallible individuals, the liberal dream of powerful institutions protecting and perfecting our lives can easily become a nightmare of corruption and abuse.

The Founding Fathers knew this. They didn't trust institutions. They didn't think a nation could remain free unless its citizens stayed on top of things themselves; that's why they set up such an elaborate balance of competing interests, of checks and balances. I believe the right approach is to reduce the power of institutions, and increase the power of the people, who have the most at stake.

The EPA Fails to Assist States' Environmental Efforts

by Kathy Prosser

About the author: *Kathy Prosser is the commissioner of Indiana's Department of Environmental Management.*

I run a state agency charged with the protection of Indiana's environment through the administration of federal and state environmental laws. On the surface, some of what I will talk about may sound like management. And it is. Because management problems are symptoms of policy illness.

As the head of a state agency, the subject is of critical importance to me. It is grounded in the realities my agency deals with every day.

Well, there's good news and bad news about the federal-state relationship, too.

The good news is that we have a federal agency to protect our environment—the Environmental Protection Agency. Early on, EPA developed rules and regulations, and established programs to implement the Clean Air Act, Clean Water Act, RCRA, etc. Then, they delegated to the states the responsibility to carry out these programs, and they sent along lots of new money to help us do it.

EPA Must Help the States

So, that's the good news. That was back in the 1970's. The bad news is that EPA hasn't changed much since then, despite the fact that the *economic* environment, the *business* environment, the *regulatory* environment, and the *public* environment have all changed dramatically.

We live in a different world than the world of 20 years ago—or last year, for that matter. The "New Federalism" of the last decade has forced state governments to shoulder a greater burden of federal programs. Now, 20 years after the birth of EPA, states are running almost all of the federal programs. So, the lead

From Kathy Prosser's speech to the American Law Institute and the American Bar Association, Washington, D.C., July 1, 1992.

has changed. The leader has become a follower. The tail is wagging the dog.

States need something different from EPA now. We no longer need a huge federal bureaucracy to oversee us. What we need is a federal agency to assist us in running programs and operations. We don't need EPA to tell us *what to do*, or *how to do it*. We need them to *help us get it done*. We shouldn't be fighting the war against bureaucracy. We need to be winning the war against pollution. To do that, some basic changes in federal policy must take place.

"We need skilled staff with experience in protecting the environment."

First, states need more people to implement programs. The number of federal mandates keeps increasing, but the number of staff available in the states is static at best. We need skilled staff with experience in protecting the environment. Finding skilled people is tough enough. Getting them to work for state salaries is tougher yet. In Indiana, we've been searching for a Director of our Air Program, and can't even find candidates to interview.

There is one easy answer to this problem. EPA should send us some of their federal employees. They ought to move these people from the front office, to the front lines . . . in the states. In Northwest Indiana, for example, we have 20 full-time environmental staff working on enforcement, inspections and public awareness issues in this heavily polluted area of the state. EPA's Region V, which includes Indiana, has 90 or more full-time staff who work on the same issues in Northwest Indiana. They have 90, we have 20. That's some oversight!

Don't get me wrong. I appreciate the Region's commitment to helping us clean up this area. But their help would be more useful if those 90 staffers worked more closely with the state, *in the state*. Not in some office in downtown Chicago.

More People and Money Are Needed

The federal government has a program now that allows some EPA employees to work in the state. It's called an Inter-Personnel Agreement. Under this program, federal staff remain on the federal payroll, but work in the state for a period of time. A small portion of their salary may be picked up by the state. The one person we have under this program in Indiana is invaluable to us. But we need more. Not only is this person familiar with the environmental programs, she knows how to move things through EPA. And this is no small task. Working side-by-side, we can better see each other's point of view, and that understanding goes a long way toward building partnerships which sustain us when there are pressures and strains between the EPA and the states.

So we need lots more people in the state programs. This requires a shift in policy and paradigm—from oversight and technical assistance—to real partnership in effort and expectation.

Second, states need more money. While mandates continue to increase, the proportion of federal money passed through to the states is dropping. In 1987, for example, EPA supported 43 percent of my Agency's budget. Today, they support only 18 percent. This downward trend in funding is bad enough, but when you combine it with state budget deficits, cutbacks, and layoffs around the nation, the impact is severe.

EPA must look for new ways to channel money to the states. We just can't keep running programs without resources. Almost half the states, for example, are thinking of returning drinking water programs to EPA because they don't have the money or personnel to implement them. I understand that California actually has returned this program.

If returning federal programs becomes commonplace, we will once again have the federal government, *layers removed* from the needs of people at the state and local level, running environmental programs. I don't think that serves anyone well.

Reexamining Policies

Ironically, along with the federal funding gap, states are severely restricted in how they can spend that dwindling pot. We need greater flexibility to spend the money we *do* get from EPA *on state priorities.* Under the current system, EPA directs the specific use of every federal dollar spent by states. EPA as a parent needs to recognize that state agencies and state leadership are mature enough and tough enough to manage the priorities and programs. They need to shift from paternalism to partnership.

"EPA must look for new ways to channel money to the states."

As states are being forced to reassess, realign and streamline, so, too, should EPA. We have a unique opportunity to reexamine what it is we're doing—to *stop doing* things that don't matter, or don't work, and focus our energy and scarce resources on those things that do.

Two opportunities come to mind. First, the management structure at EPA drives policy. EPA is separated into Divisions. There is an Air Division, Water Division, and Waste Division. And state environmental agencies are mostly set up in the very same way. The word Division describes well how this structure works, or doesn't. These programs might as well have walls between them for all the communication that goes on. The result is that air people don't talk to water people, and waste people don't talk to air people, and so on.

Yet, most environmental problems are interrelated and cross over into more than one area, or media. Rather than working on one specific problem at a time, we need a multi-media approach. Let's take a more comprehensive look at the kinds of environmental challenges we have, and then coordinate our staff and money to solve the problem. This sounds like management, but it's policy.

EPA is working hard to create a new system—one that forces coordination among the Divisions. We're implementing a pilot multi-media project with EPA in Northwest Indiana. There are many kinks to be worked out, but I have no doubt that with strong leadership from Region V, this new approach will reap big benefits for us.

Another opportunity is to rethink beans. What are beans, you ask? In actuality, beans is an unofficial EPA term describing the number of inspections conducted, permit reviews completed or enforcement actions taken by the state. For every action a state takes, we get one bean.

Regardless of the environmental impact of the action, we only get credit for one action. EPA uses these beans to evaluate success. They measure success by the number of actions we take at the state level. The more actions we take, the more beans we get. The more beans, the more successful EPA says we are. The more successful we are, the more money we get from them.

EPA developed its bean accountability system in the early 80's—not exactly the renaissance era of environmental policy—in an effort to regain earlier momentum. Each region was given a target number of beans which the states had to produce. To ensure that states produce beans, and the regions meet their targets, federal funding to states is tied to very narrow, very specific grant requirements. In Indiana's water program alone, we manage 43 different federal grants—43 just in the water program.

Focus on Quality

Nowhere in this equation is there a measurement of how environmentally protective the actions are, or how tough the negotiation is to bring closure to an issue. *Beans measure quantity, not quality.*

This bean approach limits flexibility, too. It doesn't take a rocket scientist, or an environmental lawyer for that matter, to figure out that the incentive for states is to make sure we get all our beans so we can keep getting the money to run our programs. Because states are under enormous financial pressure, we are forced to look for the easiest way to get those beans.

"[EPA] programs might as well have walls between them for all the communication that goes on."

This is not the way to set environmental priorities. Doing it this way means that the big issues, the *real* pollution challenges, play second fiddle to these darned beans. . . . But change comes slowly in any organization. It took me almost 90 days just to convince my own staff that we really *could* get business cards printed on recycled paper!

EPA now advocates a new approach, using comparative risk analysis. The idea is to compare environmental risks, and spend money on areas that present the greatest risk.

This approach is somewhat controversial since the severity of the risk is often defined by where you sit. For instance, managing municipal solid waste would likely score near the bottom of a comparative risk analysis. Yet those of us in the states know what a high priority it is at home.

> ***"Indiana and other states have been struggling with long-haul waste."***

Since 1989, Indiana and other states have been struggling with long-haul waste. EPA headquarters says solid waste is a state issue. Well, it's a state issue, allright. A 50-state issue. When almost a quarter of the trash filling your landfills comes from other parts of the country, you know the problem goes beyond your state's border. And when you are prevented from taking action by clauses in the U.S. Constitution, it becomes crystal clear that this demands national attention.

EPA's decision to stay on the sidelines has caused states to waste valuable resources struggling with insufficient solutions. Indiana has already been through two major federal lawsuits, fighting to protect our citizens. Those lawsuits weren't cheap, either in staff time or state resources. Some early leadership from EPA might have prevented this particular kind of waste.

If the bad news is that the paternalistic relationship doesn't work anymore, the good news is that a new relationship—one of partnership—*is* being forged—led by state innovations. In Indiana, for example, we have developed alternative funding sources, designed new regulatory tools, and forged new partnerships for environmental integrity. . . .

Innovative Laws

In addition to creative funding strategies, states are passing innovative legislation to close or tighten federal loopholes. For example, the federal superfund law and RCRA exclude petroleum contamination sources, such as refineries, from cleanup requirements.

Whiting, Indiana, is home to the largest inland refinery in the country. It is estimated that there are 8' of free floating oil on top of the groundwater in this community. To address this problem, legislation was passed in Indiana, giving my Agency the authority to order cleanup of these types of sites. We are only the 7th state in the nation to close this dangerous loophole.

A bill [passed] in our General Assembly makes Indiana a model in terms of the way we approach cleanup of contaminated areas. The bill allows those corporations interested in voluntarily cleaning up one of our 1,400 non-superfund sites, to do so in cooperation with our agency.

Non-superfund cleanups are a problem for companies because they get no guidance as to what level of remediation would be acceptable to the state. Traditional federal and state cleanup programs are focused entirely on the most serious sites, with no resources being devoted to working with companies wishing to do voluntary cleanups. This new law, if passed, would allow companies to work cooperatively with our agency to ensure that any amount of voluntary efforts they make is not wasted.

In the area of municipal solid waste, there are no loopholes to tighten or close, because there are no federal laws related to the management of municipal solid waste. States have had to adopt their own laws to deal with this issue. We have passed laws to manage our own solid waste, and we have acted to prohibit illegal waste from coming into Indiana. . . .

We need uniform standards. Does it really make sense, for instance, that Indiana's water quality standards for the Ohio River are tough downstream if, upstream, Ohio's aren't? Likewise, does it make sense that Great Lakes states offer different levels of protection for the world's greatest fresh water resource? The good news here is that states are working with EPA, in an attempt to agree on uniform standards for the Great Lakes. I don't know if we'll succeed, but at least we're trying.

Joining Forces

The state of Indiana has joined forces with EPA Region V to address a long-standing environmental problem. EPA calls it their geographic initiative—we just call it NW Indiana. As home to the nation's major steel and petroleum industry, this area is widely considered to be one of the most environmentally devastated areas of the country, and we need all the help we can get. By sharing staff and resources, this partnership with Region V can bring to bear maximum enforcement and cleanup in the least amount of time.

"States are passing innovative legislation to close or tighten federal loopholes."

Beyond this regional, neighbor-to-neighbor strategy, Governor Evan Bayh has turned to the very states where significant amounts of long-haul waste originate. He developed compacts with Governor Mario Cuomo of New York and Governor James Florio of New Jersey in an attempt to ensure that the laws of all concerned states are strictly enforced. These innovative compacts really work. We get the power to stop illegal waste from coming into Indiana. They get information they need to enforce their own laws.

For example, we identified 14 companies in New Jersey who were shipping waste to Indiana without the proper New Jersey permits. We stopped them from dumping, and New Jersey took enforcement action against them for operating without a state permit. After several years fighting one another, we're working together and making real progress.

We have also turned *conflict into cooperation* with the community we regulate. Most states struggle with making permit decisions in a timely fashion, and Indiana is no exception. This causes friction with the regulated community. In these tough economic times, it's in everyone's best interest to make sure that economic growth takes place in a well-protected environment.

In Indiana, we are working together, *turning adversaries into allies*. For the first time in the history of our state, we have asked business and industry to share with us their expertise to help us improve our permitting operation.

State innovations are catching on. EPA needs to catch up. Problems still exist, but the citizens of this country will keep our feet to the fire. They will not allow us to set aside the progress made to protect our earth.

I've talked about good news, and I've talked about bad news, but perhaps the *best* news is that the environment as an issue is here to stay. Some years ago, speaking about the environment, John F. Kennedy warned us that: "The nation's battle to preserve the common estate is far from won. The crises we will face may be quiet, but it is urgent."

Well, Kennedy was right. The environmental crisis *is* urgent. But it's no longer quiet. Kennedy was also correct when he said that our task is not to fix the blame for the past. It is to set the course for the future.

The EPA Wastes Taxpayers' Money

by Thomas W. Orme

About the author: *Thomas W. Orme holds a Ph.D. in microbiology and is a biotechnology consultant. Orme is the Washington, D.C., representative for the American Council on Science and Health, a New York City organization that aims to allay public fears concerning air and water pollution, pesticides, and substances considered to be toxic.*

The Superfund is America's program to clean up old hazardous waste sites. Its budget is $1.7 billion per year, approximately the same size as the National Cancer Institute's R&D [research and development] budget ($1.8 billion) and more than twice what the National Institutes of Health are spending on AIDS ($742 million). In addition, the Environmental Protection Agency has authority to order cleanups that cost American industry another two to three billion annually. This expense, which environmentalists champion as what "polluters must pay," is passed onto consumers in the form of higher prices. Since 1980, about $16 billion has been spent on cleanups. If the program is allowed to continue, about 2000 sites will be cleaned up at a cost of $40 million per site, or $80 billion conservatively.

Money for Nothing

What are American consumers getting for their dollars? Not much. The Superfund program is an expensive dump site beautification exercise that has miniscule, if any, impact on improving public health. While $80 billion is being spent to "prevent" premature disease and death from dump site chemical exposure (the actual number of deaths from this source is unknown, but the best estimate is zero), one million Americans are dying annually from known and preventable diseases, including those caused by cigarette smoking, other illicit drug use, sexually permissive behavior, failure to use life-saving technologies

Thomas W. Orme, "Superfund: Is It Bulldozing Our Public Health Dollars?" *Priorities*, Summer 1992.

like seat belts and early detection tests and more.

How were the intentions of Superfund, to protect human health and the environment from hazards in old waste sites, thwarted? This viewpoint identifies the three people who more than any others have created the Superfund scandal.

- Devra Lee Davis is the National Research Council's dump site alarmist who believes industrialization itself is an important cause of cancer. She suggested, without scientific evidence, that hazardous waste sites are jeopardizing the health of millions of Americans.
- Joel S. Hirschhorn is the former Office of Technology Assessment [OTA] official who delighted Superfund cleanup contractors with his cost estimate for the program. He suggested that dump site cleanup will require an investment of $500 billion.
- William K. Reilly is the EPA Administrator who has given a higher priority to litigation than to cleanup so that the Superfund now threatens public health rather than protects it.

These ideas are behind changes in the Superfund that divert it from its original intent and turn it into a cumbersome bureaucracy focused on extracting money from industry to sustain a program of overzealous abatement of hypothetical risks.

The Cancer Myth

The EPA estimates that 21 million people live within three miles of a hazardous waste site. The more notorious sites, such as Love Canal in Niagara Falls or Times Beach in St. Louis, have been studied extensively. Investigators tried to determine if residents living near these sites have a higher than expected frequency of cancer. They found no convincing evidence for a cause-and-effect relationship. The National Cancer Institute cannot identify a single person whose cancer resulted from exposure to waste site carcinogens. The EPA has no such evidence, nor does the Agency for Toxic Substances and Disease Registry. Nonetheless, because of the uncertainties associated with epidemiological studies, the EPA's Science Advisory Board estimated that dump site carcinogens cause 1000 cases of cancer per year. This is pure speculation.

The internationally known epidemiologists Sir Richard Doll and Richard Peto have attributed cancer deaths to various factors. The best estimate of the contribution of cigarette smoking and other forms of tobacco use to cancer is 30 percent. Unknown dietary factors contribute approximately 35 percent to cancer deaths. On the same scale all forms of pollution combined contribute about two percent to cancer death. The Superfund relates to a tiny fraction of this pollution, only old hazardous waste sites which have not been

> ***"The Superfund now threatens public health rather than protects it."***

proven to cause cancer. Yet the program attracts more government dollars than all other Federal research efforts to control cancer combined.

The National Cancer Institute studies the cancer patient and tries to determine the cause of the disease. Devra Lee Davis is looking at the problem in reverse. She examines information about man-made chemicals in the environment, then tries to find an adverse health effect she can blame on them. With the ability to measure pollutants at the level of one part per billion, it is not difficult to find some level of environmental contamination at a hazardous waste site. Trace levels of contaminants in groundwater are especially easy to find. Combined with the EPA's risk assessment methods that overestimate hypothetical risk, a plausible, but unconvincing, case is made that hazardous waste sites are a public health problem. Scientists everywhere are beginning to wake up. Where is the clinical disease? Where are all the patients?

Dump Site Alarmists

There is no sign that people go out of their way to expose themselves to toxic wastes in dump sites. No death certificates say that exposure to toxic waste in dump sites was a cause of death. To Davis' credit she has stuck with the cancer endpoint until recently trying to resist abandoning cancer, as other dump site alarmists have, in favor of more diffuse endpoints like "20th Century Disease," "Chemical AIDS," "environmental illness" and more recently, gender-linked behavior change. In this, the alarmists join the clinical ecologists who have enriched themselves by practicing a suspect brand of medicine best described as New Age quackery. Their diagnostic procedures are not scientifically sound. Hazardous waste sites are not a public health threat worth billions and billions of dollars. This idea has lost its credibility.

> ***"No death certificates say that exposure to toxic waste in dump sites was a cause of death."***

In 1989 when Administrator Reilly addressed the Natural Resources Defense Council, he alluded to the thrill of litigation:

> I am, as of this evening, a named defendant in 489 lawsuits; that number may be a record high for the agency. . . . I want to be involved in more lawsuits still—as initiator and as plaintiff.

Reilly's wish is coming true. As a result of his "enforcement first" policy, the Superfund is becoming less a program for cleaning up hazardous waste sites and more a futile legal exercise of government against industry. In 1991 the EPA spent $173 million on enforcement to obtain $1.4 billion in cleanup commitments. That, maintains Reilly, is an eight to one return on the litigation dollar. But the figure does not include litigation costs to the responsible party. Furthermore, it does not assure that any of the $1.4 billion collected will correct dump site problems.

In testimony before the Subcommittee on Investigation and Oversight of the House Public Works and Transportation Committee, Reilly defended his "successes." In 1989 and 1990, responsible party participation in site cleanup increased from 30 to 60 percent. Since 1980 over $4 billion has been obtained in commitments from responsible parties. Cleanup orders were issued at 13 sites in 1988, 28 in 1989 and 44 in 1990. Says Reilly, "In 1990, 151 Superfund cases were filed: the highest in our history and a 50 percent increase over the previous year. I expect this level of success to continue in 1991 and future years." However, the initiation of litigation should not be considered success. Health or environmental benefit resulting from cleanup would be the best measure of Superfund's success.

"Health or environmental benefit resulting from cleanup would be the best measure of Superfund's success."

Superfund's Coffers

Meanwhile, the chemical industry, a major segment of the responsible party population, has been paying special excise taxes into the Superfund Trust Account. As of September 30, 1991, the Superfund Trust Account had an equity balance of $3.15 billion, approaching the statutory limit of $3.5 billion. That represents a surplus of almost 2.44 years of Superfund tax receipts. Reilly's management of the funds already at his disposal has not been very good. Furthermore, he may not know what to do with the additional funds brought in by "enforcement first." Some $3.15 billion in the Superfund Trust Account remains idle while the EPA sues the companies that paid into it.

The effectiveness of "enforcement first" is diminishing as courts take a harder look at the costs the EPA is imposing on the chemical industry. Keith Schneider writing for the *New York Times* has reviewed moves by federal courts to reverse EPA policy and insist on cost effectiveness. He predicts that in the future EPA funds expended on enforcement will result not in money for cleanup but in more lost cases. That's not good news for the taxpayer. As the mood in the chemical industry shifts from compliance to confrontation, it will become evident that the "deep pockets" of America's bigger chemical companies can be used just as effectively to prolong litigation as they can to comply with the spirit of the cleanup effort.

This is precisely the posture Congress hoped to avoid in the carefully worded 1986 Superfund reauthorization bill. Reilly has frustrated the intention of Congress, which sought to reduce litigation and give responsible parties some control over the costs of cleanup. Those amendments allowed responsible parties to conduct their own risk assessment, specify a remedial strategy for EPA approval and direct the cleanup effort. Trust fund monies were to be used to reduce the cost to responsible parties, and minor waste contributors could make small cash payments to absolve themselves of further responsibility. Such efforts to reduce litigation and bureaucracy costs have been stifled.

Industry resistance to government-imposed costs might be expected under any circumstances, but the *coup de grace* for Reilly's "enforcement first" policy came from within EPA. Don Clay, Reilly's Assistant Administrator, Office of Solid Waste and Emergency Response, in reply to President [George] Bush's request that all agencies study how to reduce regulation that is throttling the economy and impeding recovery, suggested that it would be relatively easy to save $1 billion per year on hazardous waste program expenditures including Superfund by shoring up management and focusing on real rather than hypothetical risks. The chemical industry has greeted his suggestions enthusiastically. The taxpayer is left to wonder why such excesses have been permitted since the program's inception.

A Turn Toward Confusion

The Superfund program lost its way in 1988 when it responded to criticism contained in an OTA report written by Joel S. Hirschhorn. The Superfund had a lackluster history of slow progress to that date, but in 1988 it took a decisive turn toward bureaucratic confusion and away from focused environmental restoration. CERCLA (the Comprehensive Environmental Response, Compensation, and Liability Act), the legislation that authorized the Superfund, became law under President Jimmy Carter in 1980. It received lukewarm attention from the incoming Reagan administration and languished for two years from 1984 to 1985 while reauthorization was considered. SARA (Superfund Amendments and Reauthorization Act), the 1986 reauthorization of CERCLA, was a real achievement in non-partisan objectivity that allocated $8.5 billion to cleanup efforts and addressed responsible party financial issues with fairness. It forged a spirit of cooperation between government and industry, necessary for success because of the *ex post facto* nature of the liability assignments it authorized. In 1988, however, Mr. Hirschhorn disclosed to a House Oversight Committee preliminary findings of a report, "Coming Clean: Superfund Problems Can Be Solved."

> ***"The Superfund . . . took a decisive turn toward bureaucratic confusion and away from focused environmental restoration."***

Mr. Hirschhorn and his staff had examined Records of Decision, documents that described the rationale for adopting a particular technology to clean up a hazardous waste site. These documents have significance because they are prepared after all preliminary investigations have been completed and are discussed publicly as part of "citizen participation."

Mr. Hirschhorn found that cleanups financed and directed by responsible parties were on the average costing less than those financed and directed by the government. This finding would not have surprised Nobel Prize winning economist Milton Friedman or Senator Warren Rudman (R-NH), both critics of the Federal government's wasteful spending practices. It suggested to Mr. Hirschhorn, however, not that government was spending too much but that responsible parties were not spending enough. Mr. Hirschhorn proceeded to obscure the distinction between a solid waste site and a hazardous waste site, confused the long-term objectives of the Resource Conservation and Recovery Act with the short-term objectives of the Superfund and introduced $500 billion as a realistic cost for permanent cleanup of hazardous sites. This is a very high cost for a program funded at a level of $2-3 billion per year. Furthermore, Hirschhorn reported personnel shortages and deficiencies in technology. His solution was a massive expansion of government. In Hirschhorn's estimation, industry was not trustworthy. The EPA should make all site evaluations, should devise and implement cleanup strategies and should sue the chemical industry before, during and after to pay the costs. Above all, cleanups had to be "permanent." Hazardous waste should be destroyed, not contained or consolidated and transferred to other sites. A cleanup should result in a pristine environment, offering what the Chemical Manufacturers Association later would ridicule as "edible soil and drinkable leachate."

"Litigation is not the answer. 'Enforcement first' is not working."

Redirecting Superfund

Can Superfund get back on track? Yes, but not without major restructuring. The EPA has shown that waste sites can be secured at a small cost. Exposure to high doses of hazardous chemicals then becomes unlikely. Litigation is not the answer. "Enforcement first" is not working. The spirit of 1986 when industry and government cooperated on cleanup goals and finances can be recaptured if cost estimates for cleanup are reasonable. New concepts in large scale recycling-incinerator-power plants are a key to success. Their construction addresses not only the cleanup problems of old dump sites, but also resolves the coming landfill crisis.

Hazardous waste sites represent only one of many public health problems confronting Americans. Each problem has a priority. Assigning the correct priority is a responsibility of the President and the Congress. The National Research Council, the EPA and the Office of Technology Assessment . . . have failed in their assignment of priority. Their advisors and peers must put the Superfund program into perspective and redirect the wasted dollars toward activities that are more likely to enhance public health.

EPA Regulations Oppress Small Manufacturers

by Robert M. Cox Jr.

About the author: *Robert M. Cox Jr. is an industrial paints expert who volunteers to help clean up waste sites. He is the former owner of a paint company that was forced into bankruptcy because of harsh EPA regulations.*

I am the former owner and president of the Gilbert Spruance Co., a Pennsylvania manufacturer of industrial coatings for wood cabinets and furniture. I was president from 1986 until May 1992. I purchased the company from my grandfather's estate, in November 1989. Until December of 1991, I owned the company running it as a "debtor in possession," although never declaring formal bankruptcy.

In 1984, we at Spruance were notified that our former waste hauler, Mr. Marvin Jonas from New Jersey, testified under implied immunity to the EPA that he had hauled waste for Du Pont, Hercules Chemical, Columbia Broadcasting Systems, Texaco, and Gilbert Spruance, etc.

The Superfund Juggernaut

We had been a small- to medium-sized family paint company since 1906, doing about $5,500,000 in annual sales. Once Jonas testified, under Superfund and RCRA laws, Spruance was considered to be a Potentially Responsible Party or PRP in all Jonas sites, and a nightmare of epic proportions developed at Spruance. We found ourselves under the "Superfund Juggernaut," with no means of escape.

The Superfund law carries "joint and several liability," along with the threat of treble damages. "Joint and several liability" means that once you are identified as a PRP you can be held responsible for the disposal, in our case, of all of the Jonas waste in every site he used in New Jersey, with the threat of damages being tripled if one does not comply with waste site cleanup directives. Jonas

From Robert M. Cox Jr., "Superfund: One Company's Nightmare," position paper written expressly for use in the present volume.

testified about all the sites he used in New Jersey to dump waste which, it turns out, he did illegally. We at Spruance had no way of knowing this at the time, since in the 1960's and 70's Jonas showed all proper paper work approved for the hauling of waste.

With Jonas using ten to eleven sites in New Jersey, we at Spruance were implicated in the dumping at these sites. It did not matter that our records did not show our waste going to these sites, it only mattered that we used a "gypsy hauler" named Jonas. We were an extremely deminimus contributor. The Superfund law makes an effort to define deminimus as a small contributor, but is not specific as to the course deminimus contributors should take. Generally speaking, deminimus is less than one percent of the total waste in the site, but usually arguments develop in liaison counsels as to who should pay for what and how much. The "battle lines" are drawn between the larger companies against the deminimus smaller companies. All of this arguing goes on amongst lawyers, consultants, and administrative types, at the rate of $150-$300 per hour. But after settling one site for more than $200,000, including legal fees, we were tiered in other sites along with Texaco and Rohm & Haas, a large Philadelphia supplier, and asked to pay anywhere from $175,000 to $1,300,000 to get out of our Superfund liability *per site*.

"We found ourselves under the 'Superfund Juggernaut,' with no means of escape."

Futile Efforts

Spruance had no choice but to fight the situation. Our insurance company Pennsylvania Manufacturers Association (PMA) abandoned us after the $200,000 settlement. In 1987 they won a declaratory judgment so that they no longer had to pay legal costs or indemnify any Superfund issues involving Spruance.

We at Spruance countersued in New Jersey. We lost at the lower court, but achieved a favorable ruling at the appellate level. PMA then decided to take it to the Superior Court of New Jersey. As of May 1993 there has been no decision.

Meanwhile, since 1985 Spruance paid approximately $250,000 in legal fees in defense of Superfund, plus one-third of our executive staffs' time with nothing to show for our efforts.

We tried pleading to aides of Senators Bill Bradley, Frank Lautenberg, and Arlen Specter the injustice of this issue, but to no avail. Our National Paint and Coatings Association along with others tried to get Congress and our government to see the inequities of the Superfund law. Nothing has happened to this moment.

Meanwhile, lawyers and administrative consultants keep charging Spruance for their involvement in Superfund liaison counsels and third party suits. Administrative and consulting fees involving Superfund sites can range as high as

$8.00-$12.00 per gallon of waste. Liaison counsels remind me of a "high stakes poker game" where, as a PRP, you are asked to ante up anywhere from $5,000 to $15,000 to be told that you now owe anywhere from $175,000 to $1,300,000 to absolve your liability in the site. Even then, there is a possibility you can become involved again at a later date, depending upon site progress in remediation and cleanup.

A Drain on Business

We at Spruance lost focus on our *core* business; we felt like we were under total "Superfund attack." We had to add between $.50 to $1.10 per gallon of coating to cover just our legal costs. We also needed to develop contingent liability provisions in our financial statements to show the true value of our company. Our customers did not understand our total predicament, and our products became less competitive in price, the more involved we became in Superfund. It was impossible to budget for long-term expenditures in research and development and in proper training for our work force to meet the challenges of the future, with Superfund site charges and legal fees "rocking us," in many cases as a surprisingly high figure. I was receiving at least two times per week stacks of mail involving Superfund at Spruance. We had more files on Superfund than we had anything else.

> ***"We at Spruance lost focus on our*** core ***business; we felt like we were under total 'Superfund attack.'"***

Our choice was to declare bankruptcy, or sell Spruance, since we were running out of cash. No bank in Philadelphia would lend any working capital to a company beset with environmental problems like Spruance.

We tried to settle with EPA. And we tried settling with local DERS (Departments of Environmental Resources for New Jersey). Nothing happened except the continued onslaught of legal bills. I'm convinced the legal profession and large companies' mission here was to prolong the Superfund issue to obtain their fees; and for the large companies and insurance companies to put small companies under.

I finally sold the company in December 1991 to a multi-state purported public firm named Capital Pacific. They compounded the problem by not paying any taxes after promising a working capital influx of $300,000. I was forced to resign 10 months after selling the company, which was eventually seized by the IRS, and still the inertia continues.

A Stand Against Injustice

Now I'm a "baby boomer," who grew up in the 60's and very much supported what Robert F. Kennedy stood for, trying to keep these principles alive. I'm standing up for a gross injustice. As Senator Kennedy stated: "Anyone who stands up for an injustice, provides a ripple of hope throughout the entire country."

I was born on April 22, 1947, which is now recognized as Earth Day. I've attended Earth Day celebrations since its first on April 22, 1970. I am basically an environmentalist, who felt an industrial coatings company concerned with the environment could work with our environmental agencies. I'm not going to give up hope. I'm an optimist by nature, and I want to apply some reasonableness to the Superfund process. Why doesn't the system allow an environmentally sensitive paint company to co-exist with Superfund? The intentions of both are similar. Why does our system create so much inertia that the lawyers and administrators get rich off a manufacturing company trying to do the correct thing all along?

How can our legislatures, executive branches, and judiciary system allow a hauler named Marvin Jonas to live in Costa Rica and not pay the proper penalty for his wrongdoing? I feel that my due process and our company's due process was violated. I will continue to stand up for this "injustice" until our "ripple of hope" meets with someone who cares about the small companies' plight with this unjust law.

Put People First

A Superfund nightmare became a Spruance tragedy, but it does not have to be for other small companies facing this scenario. My hope continues to be that our administration, with President Bill Clinton and Vice President Al Gore, will be able to balance environmental concerns with small business needs. That the promise of this new administration does not get circumvented by a Superfund law that continues to chew up companies, and not allow our country to reestablish our proper manufacturing base. Spruance is like Rhode Island at our First Constitutional Convention. Our founding fathers made proper allowances in our Constitution to allow Rhode Island to be "heard" and have proper stature. I'm sure we can allow the same guidelines to be used now so the tragedy of Spruance is not repeated. Let's be sure "to keep people first," and allow our freedoms to *not* be buried under bureaucratic red tape, allowing the few to prosper from a system that needs fixing.

Chapter 4

Is Recycling an Effective Way to Reduce Pollution?

Recycling: An Overview

by Bruce van Voorst

About the author: *Bruce van Voorst is a senior correspondent for* Time, *a weekly newsmagazine.*

It's a self-congratulatory ritual, repeated every day, every week, all over America. Separate the clear glass bottles from the green and amber ones. Place the newsprint in one basket, mixed white paper in another, the reams of used computer paper in a third. Haul the whole lot out to the curb. There. You've just done your bit for humanity: you've recycled. It's Miller time.

Not so fast.

To be sure, recycling is in vogue. Citizen participation is at an all-time high; curbside collection programs have exploded from 600 in 1989 to 4,000 today. But the dirty secret, and it's not a little one, is that major quantities of the material being collected never actually get recycled. More than 10,000 tons of old newspapers have piled up in waterfront warehouses in New Jersey, and a congressional committee has heard testimony that the nationwide figure tops 100 million tons. At the Pentagon, employees looking out over the parking lot can watch paper they've carefully segregated in the office being tossed into a single dumpster, destined for an incinerator. The used-glass market has been so soft that Waste Management of Seattle, Inc. is stuck with a mini-mountain of 6,000 tons of bottles from neighborhood collections. In the Minneapolis-St. Paul area, haulers have run out of storage space and are incinerating some recyclable goods. "It's like having your suitcase all packed with no place to go," laments Amy Perry, solid-waste program director for the nonprofit Massachusetts Public Interest Research Group.

The problem is that the economics of recycling are out of whack. Enthusiasm for collecting recyclables has raced ahead of the capacity in many areas to process and market them. Right now, says Victor Bell, a veteran Rhode Island recy-

Bruce van Voorst, "The Recycling Bottleneck," *Time*, September 14, 1992.

cling expert, "the market can't keep up with the recycling binge." In recent years many states and municipalities have passed laws mandating the collection of newspapers, plastics, glass and paper. But arranging for processing— and finding a profit in it—has proved tricky. As trucks loaded with recyclable materials arrive at processors, backlogs develop. Worse, the glut has depressed already soft prices for used paper and plastics.

> ***"The problem is that the economics of recycling are out of whack."***

"Long term, our members recognize that if you're not in recycling, you'll be out of business in 10 years," says Allen Blakey, public relations director for the National Solid Wastes Management Association, the nation's trash collectors. Yet government-mandated recycling laws, by requiring haulers in some instances to pick up unmarketable items, are actually forcing some into bankruptcy. The danger in this short-term failure of recyclonomics, warns William Rathje, author of the recently published book *Rubbish! The Archaeology of Garbage*, "is that, in the interim, recycling enthusiasts will become disillusioned at reports of difficulties." If there's money in trash, entrepreneurs will find it. And in many instances they have. Processors are turning a profit by recycling high-value steel and aluminum cans and, in general, paper cartons and cardboard. A Shearson Lehman analysis concludes that recycling is now attracting "the attention of the solid waste industry investor." In two areas in particular, innovative ideas are cropping up:

NEWSPRINT. Paper, especially newspaper, is the biggest component of landfills—about 40%. Despite being the most widely recycled material, newsprint is not at all easy to process or market. "Often we can't give the stuff away," says James Harvey, owner of E.L. Harvey & Sons, Inc., a Westboro, Massachusetts, hauler. Facilities to remove ink from newsprint—a necessary step before it can be pulped to make new paper—are enormously expensive. To justify the investment, recyclers need the sort of arrangement just announced between the city of Houston and Champion Recycling Corp. In return for building an $85 million de-inking plant, Champion Recycling, a subsidiary of Champion International Corp., a leading paper manufacturer, was assured of getting the city's entire collection of old newspapers and magazines. "Our customers not only want to buy recycled materials; they are insisting on it," says Champion International president Andrew Sigler. "This is a market-driven operation that's great for Houston and gives us the assured supply we need for economic efficiency."

PLASTICS. Though plastics constitute 8.3% of all municipal solid wastes and are proliferating faster than any other material, less than 2% of waste plastic gets recycled. Largely this is because it is cumbersome and expensive to separate the seven basic types and relatively cheap simply to manufacture virgin plastics. Wellman Inc., of Shrewsbury, New Jersey, has emerged as a leader in recycling so-called PET [polyethylene terephthalate] bottles, the most common clear plastic containers for liquid, turning discarded ones into furniture textiles, tennis balls, electrical equipment and yarn for polyester carpet. The Coca-Cola Co. services major markets nationwide with two-liter bottles made of 25% recycled PET plastic.

"It will always cost you money to get rid of garbage," asserts Marcia Bystryn, a recycling official in New York City. The trick is to encourage behavior that minimizes the costs, allocates them as equitably as possible and creates productive economic activity wherever possible. In large measure, the present disequilibrium in recycling is the result of policies that work at cross-purposes with those goals and with one another. Environmentalists argue—correctly— that recycled materials suffer in the marketplace against virgin materials because of government subsidies. Newsprint producers, for instance, are indirectly subsidized through public-area logging and logging access roads. The depletion allowance for petroleum subsidizes producers of oil-based plastics. "If these costs are taken into consideration," contends Allen Hershkowitz, senior scientist at the Natural Resources Defense Council, "recycling looks economically a lot more competitive."

"If there's money in trash, entrepreneurs will find it."

Even with such disadvantages, there are profitable recycling operations. Three years ago, J.J. Hoyt, recycling manager at the U.S. Naval Base in Norfolk, Virginia, took over a solid-waste disposal program that had been costing taxpayers $1 million a year. A shrewd businessman, Hoyt was sensitive to hauling managers' needs and negotiated lucrative deals. Now, says one Navy officer, "not a tin can or newspaper falls to the ground on base." This year Hoyt's program is earning close to $800,000. "The key is knowing the market," he says.

New York City's experience is decidedly more mixed. Its primary landfill, Fresh Kills on Staten Island, already covering 2,200 acres and rising to a height of 155 ft., is rapidly filling up. And the city, which recycles only about 6% of its waste, must turn increasingly to recycling or incineration. A program launched in 1989 to recycle 25% of the city's daily output of 26,000 tons of solid waste has fallen short. Only 29 of the city's 59 community board districts participate in the program. Although Mayor David Dinkins hopes to expand this to 39 by the end of the year, officials admit that recycling faces heavy slogging. "Recycling began with a real naive sort of optimism," says Bystryn. "I think it is important to come back somewhere near to reality." The Dinkins administration succeeded against intense environmentalist opposition in enacting a waste-disposal plan that includes construction of an incinerator in Brooklyn.

Critics of recycling in the U.S. claim that it weakens the economy, but Germany, one of the world's strongest economies, is showing that isn't necessarily so. Since last December, manufacturers and retail stores in Germany have been required to take back such transport packing materials as cardboard boxes and Styrofoam. This spring the requirement was extended to "secondary packaging" such as cardboard boxes for toothpaste or deodorants. By next year, consumers will be able to return sales packaging—from yogurt cups to meat wrappers—to the point of purchase for disposal. In mid-1995 German manufacturers will be responsible for collecting 80% of their packaging waste. Augmenting the government's program is the Duales System Deutschland, a private-industry-initiative recycling program that has already distributed collection bins to more than half of Germany's 80 million people and expects to reach virtually 100% before the end of the year.

Japan's recycling rate is more than double that of the U.S.—40% of municipal solid waste, vs. 17%. But the Japanese program shares some of the problems familiar to American recyclers. Milk cartons, one of the favorite recycling items, are piling up high in warehouses. Like America, says Hiroshi Takatsuki, a professor at Kyoto University, "Japan emphasized collection before coming up with an appropriate infrastructure for reuse."

Americans dispose of far and away more waste than anybody else on the planet. The EPA estimates the annual cost of this disposal at more than $30 billion, a figure rising 17% a year and predicted to reach $75 billion by the end of the century. On the other hand, despite the dire predictions of some environmentalists, disposal is less of a problem than in many other countries. There are still plenty of landfills available, and they will continue to play an important role. So will new incinerators, despite their many environmental shortcomings. For America to catch up in recycling, experts call for action in four areas:

ECONOMICS. Recycled materials deserve at least the same tax and subsidy treatment that is provided for virgin materials—especially paper and plastics. Potential investors in recycling equipment and research should be encouraged with tax incentives.

PACKAGING. About 39% of the paper and paperboard going into landfills and incinerators comes from packaging. The German example shows how that number can be dramatically reduced. Lever Bros., for instance, manufactures a superconcentrated powder laundry detergent in small boxes, saving the equivalent of 13 million plastic bottles a year. L & F Products sells its Lysol brand and other liquid cleaners in Smart Packs that take up 65% less landfill space than the jet-spray containers they are designed to refill. Imperial Chemical Industries of London has developed a plastic, soon to be distributed in the U.S., that biodegrades with or without exposure to air and sunlight.

> ***"Recycling began with a real naive sort of optimism. . . . I think it is important to come back somewhere near to reality."***

RESEARCH AND DEVELOPMENT. Recycling is a new frontier for technical innovation. New processes, for instance, are needed to remove contaminants. Sorted solid wastes often include contaminants that gum up recycling systems, such as clear plastic tape on envelopes or sticky yellow Post-its on office paper. A single ceramic cap from a bottle of the Dutch-brewed Grolsch beer can contaminate an entire batch of green glass. "We haven't begun to tap the potential for technical innovation in recycling," says Lloyd Leonard, legislative director for the League of Women Voters.

LEGISLATION. The New Jersey mandatory recycling law—achieving 34% recycling, or double the national average— demonstrates the virtues of a legal prescript. Minimum-content laws such as those in Oregon and California, mandating the use of recycled materials in new products, have proved effective. So have "pay by bag laws" that increase the price tag for garbage removal according to volume. Last fall the White House issued an executive order requiring federal agencies to give preference to recycled materials when purchasing products. But that's just a start. "Unless the government mandates more use of recycled material in products," warns Dan Weiss of the Sierra Club, "recycling will be discredited."

For all its promises, recycling remains only part of the world's waste-disposal solution. Despite the enormous energy and enthusiasm with which Americans and others collect recyclable products, the real breakthrough can come only when similar effort is expended on reducing waste in the first place and in enticing more markets to absorb recycled materials.

Recycling Should Be a High Priority

by Jennifer Carless

About the author: *Jennifer Carless, a free-lance writer on environmental issues, is the author of* Taking Out the Trash: A No-Nonsense Guide to Recycling, *from which this viewpoint is excerpted.*

Recycling is getting more and more publicity these days, but what exactly is it? Recycling is returning materials to their raw material components and then using these again to supplement or replace new (virgin) materials in the manufacture of a new product. The process involves several steps: separating materials from the waste stream, collecting them, processing them, and ultimately reusing them either as an entirely new product or as part of a new product. It is important to understand at the outset that a material has not been recycled simply by having been collected from a curbside or taken to a buy-back center: it is only truly recycled once it has been turned into another product.

In a more general sense, recycling also means simply putting something you were going to throw away to good use. We do this daily when we donate clothes to charitable organizations or when we reuse a plastic container to store food in the refrigerator.

Some of the more common [recycled products] include newsprint, aluminum cans, and glass. To appreciate the value of these commodities, we must stop thinking of recyclables as garbage or waste, which have negative connotations, and start thinking of them as secondary raw materials. Recyclable materials are not trash; they are resources.

Toward an Integrated Solution

Most experts believe that to control our increasing waste disposal problems in this country we need to adopt an integrated waste management approach. Although one can find diverse definitions of this concept, most agree on its general principles. An integrated approach to managing our waste involves source reduction, reuse, recycling, and then (and only then) either landfilling or incineration as a final disposal method. In other words, the goal is to eliminate as much as possible of our waste stream by either reducing the amount of waste we generate in the first place or at least reusing or recycling everything possible. While this viewpoint focuses on recycling, keep in mind that it's one part of an overall solution.

Source reduction, also called waste reduction, can be accomplished in many ways. Buying in bulk, avoiding disposable products, and buying durable (and repairable) goods are all excellent waste reduction strategies. Industry can help by reducing the size or weight of a product or its packaging. As an example, today's aluminum can is lighter than its counterpart of years ago, as is a two-liter plastic container.

Reusing things is simple once the habit is established. Using glass bottles to hold frozen juice, saving wrapping paper, taking shopping bags to the store—all are examples of reusing materials. Photocopying and writing on both sides of paper and reusing envelopes are others.

"We must stop thinking of recyclables as garbage or waste . . . and start thinking of them as secondary raw materials."

Recycling benefits everyone, both in obvious ways and in ways many people haven't even considered. The most obvious benefit to recycling is that it saves landfill space, which is certainly very important. Recycling does much more, however. It helps all of us in a multitude of subtle ways. The environment, individuals, local communities, and industry all realize concrete advantages from recycling reusable materials. Recycling is easy and provides immediate results.

Environmental Benefits of Recycling

First and foremost in most people's eyes are the environmental advantages of recycling. These are numerous. Our land, air, and water are affected by everything we do—and by everything we don't do. Recycling allows us to conserve our precious natural resources and energy, contributes significantly to a reduction in pollution, and eliminates the negative environmental impact of alternative disposal methods such as landfilling and incineration. It helps to preserve treasured wildlife habitats and vital ecosystems.

The conservation of our natural resources is certainly one of the most compelling reasons to recycle. Most of the products we use daily have been manufactured from extracted virgin materials. But by recycling paper products, glass, metals, aluminum, yard wastes, and many other materials, we can significantly reduce the demand for raw materials to produce the goods we consume. As an example, seventeen pulp trees are saved by each ton of paper made from recycled materials. In addition to saving resources, recycling helps preserve the natural landscape by requiring less mining of raw materials.

Recycling results in significant savings in other areas as well: energy, water, raw materials, and capital can all be saved by reusing what we have already produced. Approximately half the water is needed to manufacture recycled paper than is needed to produce paper from virgin pulp. Another example can be found in the aluminum industry. Alcoa, a major aluminum recycling company, states that an industrial facility to melt used aluminum cans may be built in half the time and one-tenth the cost of the facilities required to mine and refine ore to produce aluminum.

Recycling saves enormous amounts of our precious energy and the fuels which go to produce that energy because manufacturing new products from secondary, as opposed to raw, materials is typically more efficient. Our waste stream is a huge source of untapped energy that sadly, when recycling doesn't happen, literally goes to waste.

Let's take just one illustration of the energy efficiency to be gained from recycling: the production of recycled paper requires anywhere between 23 and 74 percent less energy per ton than does that of virgin paper. Another illustration: producing one aluminum can from recycled aluminum rather than from raw materials saves the energy equivalent of one-half of that can filled with gasoline. In fact, just by increasing the levels of steel and paper recycling in this country we could afford to shut down numerous nuclear power plants.

Reducing Air and Water Pollution

When recycled materials are used in the manufacturing process instead of virgin materials, there is almost always a significant reduction of harmful emissions. Both air and water pollution can be significantly reduced. Manufacturing with scrap paper, for example, results in lower levels of harmful emissions into the environment compared to the pollution that results from virgin wood pulp paper manufacture. Specifically, the EPA has found that a paper mill can reduce its air pollution by 74 percent and its water pollution by 35 percent by using wastepaper instead of virgin pulp. This means that every ton of paper made from recycled pulp keeps nearly 60 pounds of air pollutants out of the atmosphere.

"Recycling allows us to conserve our precious natural resources and energy."

Our water is affected in much the same way. Factories using raw materials pollute streams and rivers into which they dump their waste more than factories using recyclables. A steel mill, for example, can reduce its water pollution by 76 percent and its mining wastes by 97 percent using scrap metal as a feedstock rather than iron ore, according to the Institute of Scrap Recycling Industries (ISRI).

Everything we do not recycle has a chance of either contributing to our general litter problem or being dumped into the ocean. No one needs convincing of our litter problem, and much of what we see strewn on the side of the road or in our parks is paper, aluminum, or plastic—almost all recyclable. The litter in our oceans is perhaps less obvious, but nonetheless constitutes a real disaster. Plastic is a particular problem—more than 45,000 tons of plastic waste are dumped in the world's oceans every year. Six-pack rings, fishing line, and strapping bands entangle and kill seabirds, fish, and mammals in alarming numbers each year. The National Oceanic and Atmospheric Administration (NOAA) believes that approximately 30,000 northern fur seals alone die yearly from entanglement in netting.

> ***"Producing one aluminum can from recycled aluminum . . . saves the energy equivalent of one-half of that can filled with gasoline."***

Bad Alternatives: Landfills and Incinerators

Apart from the positive reasons for recycling, other incentives stem from the negative environmental impact of alternative disposal methods. Anything we do not recycle will probably find its way to an incinerator or a landfill. There are significant concerns about air pollution from incinerators that burn municipal solid waste, much of which could have been kept out and reused. Likewise, both ground and water pollution is caused when landfills leak—which older ones tend to do with frightening regularity.

Continued reliance on landfills without the serious development of safe alternatives such as recycling simply exacerbates our nation's significant problem of the shrinking availability of landfill space. For each ton of wastepaper separated from municipal solid waste, for example, ISRI notes that more than 3 cubic yards of landfill space are saved.

Recycling, coupled with source reduction, is the best way to lessen our reliance on landfills and incinerators. Although the EPA has set a goal of 25 percent of our solid waste to be recycled, some experts estimate that up to 80 percent of the household waste stream can be recycled. This shows how far we still have to go from the current 13 percent that we are recycling. Fully comprehensive recycling programs can eliminate the need for expensive and harmful incinerators and greatly reduce our reliance on the ever-shrinking supply of landfill space. . . .

There is no doubt that recycling will increase in significance as an integral part of our waste management processes in the future. We have seen ample evidence to sustain this belief: recycling offers a sound solution to many of the growing problems with our current system; the many benefits of recycling make it a logical option; and the continued creation of recycling legislation confirms its place as a solid waste disposal tool for generations to come.

We have seen recycling develop from a haphazard system of scavenging, through neighborhood recycling, to the current flourishing programs across the country with an attendant industry developing its own state-of-the-art technology. Over the past three decades our recycling rate has increased from less than 7 percent of our waste stream to today's 13 percent recycling rate.

The recycling rates for some of the major recycled products, like paper, glass, and aluminum, are gradually increasing, though plastics still lag behind. Even lesser-known recyclables, like tires, batteries, and various scrap metals, are getting more attention and hence more recycling. Legislation encouraging the development of markets for some of these less common recyclables is helping the situation.

Still, there's a great deal of room for growth. It has been demonstrated that individuals can make a difference—in home, school, and office recycling programs. Cities, counties, and states can run successful recycling programs also, whether they be mandatory or volunteer. Only a continued effort on the part of everyone concerned will ensure a satisfactory resolution to today's waste problems. . . .

The Role of Legislation

Although many people feel that legislation should be a last resort in our attempts to develop recycling, others say we are now at the point where it is indeed necessary. In many cases it looks as if nothing else will create a demand for recycled products or help find solutions to the reuse of items like tires and batteries. There are several initiatives pending. Among these are several bills to develop recycling that have been proposed by Congressman Esteban E. Torres (D-CA).

The Newsprint Recycling Incentives Act (HR 873) is designed, as its name suggests, to increase the use of recycled newsprint. The bill would require newsprint manufacturers to produce a certain percentage of product with recycled-fiber content. The recycling requirement would be set at a rate only two percentage points higher than the current recycling rate and would increase by 2 percent a year for the next ten years. Manufacturers without the capacity to use recycled fibers (or to use enough to reach the quota) could purchase "recycling credits" from those manufacturers with excess capacity. The idea is to increase the demand for collected newsprint by stimulating the demand for paper with recycled content. The bill recognizes that current manufacturing plants are at or near capacity in using recycled fiber and thus provides an incentive to increase that capacity at a steady rate.

> ***"Anything we do not recycle will probably find its way to an incinerator or a landfill."***

The Lead Battery Recycling Incentives Act (HR 870) is another example of pending legislation. Market forces currently dictate how much lead is recycled (more when virgin lead is expensive, less when it is cheaper), and the environment and our health lose out when lead is not recycled. This bill is designed to ensure lead recycling even when the price of virgin lead falls.

Working along the same lines as the newspaper recycling act, this measure would require battery manufacturers to produce batteries with a specified percentage of recycled lead. Again, the recycled percentage would begin at a rate two percentage points higher than the current rate of recycling and increase by 2 percent annually for ten years to a level of at least 95 percent.

"Continued creation of recycling legislation confirms its place as a solid waste disposal tool."

This act as well would develop a system of credits. Any manufacturer producing in excess of the amount of reclaimed lead required could sell a "lead recycling credit" to those who produce batteries with less than the specified rate of reclaimed lead. The percentage of recycled lead content is set by law, but not the price of the credit. It is proposed to let the market establish the price. The recycling credit system intends to put more money in the hands of those producing batteries with the higher percentage of recycled lead. Then, if the price of virgin lead falls, this manufacturer is cushioned, by means of his credit income, from the relative high cost of recycled lead.

The recycling credit system was first introduced in Congress in 1989 as part of the Consumer Products Recovery Act (HR 2648/S 1181). It was later incorporated into other recycling incentive measures such as those mentioned here. This system, developed by Congressman Torres, Senator John Heinz (R-PA), and Senator Tim Wirth (D-CO), was based on work begun in Project 88, commissioned by Senators Heinz and Wirth.

Success Lies Ahead

Initiatives such as these may well pave the way for the development of equitable solutions to many of the barriers currently facing recycling in this country. While legislation may not be everyone's first choice as a method to encourage recycling, many current suggestions for legislative action do appear to offer realistic approaches to the problem. As in any new situation, a period of trial and error may be expected to smooth out unforeseen problems.

The ideas presented throughout this viewpoint will no doubt be tested, adjusted, and improved upon over the coming years. New and better ideas will come along to supersede or supplement these as participation in the debate becomes more widespread.

Because the environmental and social problems facing this country from our inadequate waste disposal system are very real and very urgent, discussion and education are essential. Only through hard decision making and creative solutions will we be able to clean up some of the mess we have created for ourselves and future generations.

Recycling is an environmentally sound and practical answer to a large part of our waste disposal problems. Its widespread application will depend on as many people as possible learning about its benefits and putting recycling to work. Only by understanding our problems and the options can we become part of the solution.

A Comprehensive Recycling Strategy Is Needed

by Peter L. Grogan

About the author: *Peter L. Grogan is president of the National Recycling Coalition, an organization in Washington, D.C., that encourages recycling. Grogan is also a partner of R.W. Beck and Associates, an engineering consulting firm in Seattle.*

The United States does not have a comprehensive solid waste recycling strategy. Since the action on waste reduction has been at the state and local levels, it would be more accurate to ask if we are on the right track in our approaches (plural).

The 42 states that have recycling laws each have a different approach, but most involve target waste reduction goals that will be met primarily by source reduction, recycling, and composting.

We can analyze those recycling approaches in general terms by looking at their three common components: the collection of recyclable materials, the processing of recyclable materials, and the marketing of recyclable commodities.

Collection Services

Are our collections approaches on the right track? Some of the evidence is optimistic. More than 4,000 local governments now provide residential collection of recyclable materials, and that number grows by nearly 1,000 each year. That means we are providing waste recycling collection services to about a quarter of the population. This is a good start, but we have generally under-invested in recyclables collection programs and over-invested in collection designed for disposal.

Other elements of our collections approaches are definitely on track. For

Peter L. Grogan, "A Forum: Will the U.S. Recycling Approach Work?" *EPA Journal*, July/August 1992.

example, many of the trial-and-error practices of the recycling collection systems of the 1980s will be replaced in the 1990s with surer methods. The emergence of cost-based rates, which work much like utilities' charges for water and electrical services, is a good trend. Charges for solid waste services are thus correlated to the volume generated.

Seattle and Portland are two of the first large cities to experiment with variable-rate systems at the residential level, with notable success. In these cities the more waste you generate, the more you pay. The outcome, especially in Seattle, has been reduced waste generation, increased recycling and composting, and resultant financial savings for the ratepayer.

"We are providing waste recycling collection services to about a quarter of the population."

Another positive direction for residential collections involves the demise of the single-collection vehicles, which some cities now use to provide services for trash, recyclable materials, and organics. The next generation of collection systems—often referred to as "2-sort" or "4-sort" programs—will radically change waste collection and processing beginning in the mid-1990s.

As the standard trash-compactor truck goes the way of the rotary dial phone, the goal will become to collect all of the materials, with maximum waste recovery, in one or two vehicles. Residential and commercial sectors will participate in these collection programs, and rural areas will be served by a truck that is part compactor, part recycling truck. These vehicles are already serving communities like Telluride, Colorado. The result: higher waste recovery and lower collection expense. As this shift takes place, it will represent the first step towards appropriate capitalization of hardware.

Material Recovery Facilities

Are we on track to providing the appropriate infrastructure for processing recyclable materials? Here, again, the nation is off to a good start, but a lot of work lies ahead.

The system of choice for processing recyclable materials at the local government level is the material recovery facility, or MRF. Approximately 250 MRFs are now recovering recyclable materials from local governmental programs. The primary role of these facilities is to remove materials from the waste stream and return them to the stream of commerce.

The trend in MRFs represents an under-invested but appropriate direction that needs bolstering in two areas. First, we need to build a national MRF infrastructure, just as we have for other governmental services, such as fire protection.

The second problem area is design. Most of today's facilities were designed to process only residentially collected recyclable materials. If future facilities are set up for 2-sort and 4-sort collection programs, they will be designed to process most of the entire waste stream. Economics will be improved by the larger mission of future MRFs through economies of scale.

Recycling costs cannot be aptly compared to the costs of traditional solid waste collection and disposal simply by comparing the cost per ton of each service. A true comparative analysis must take into account, on the side of traditional disposal practices, such factors as hidden tax supports; economic impacts from environmental degradation; the costs of proper landfill closure and post-closure procedures such as those spelled out in EPA's 1991 regulations setting the first comprehensive federal standards for municipal landfills; and liability considerations.

America's Annual Solid Waste

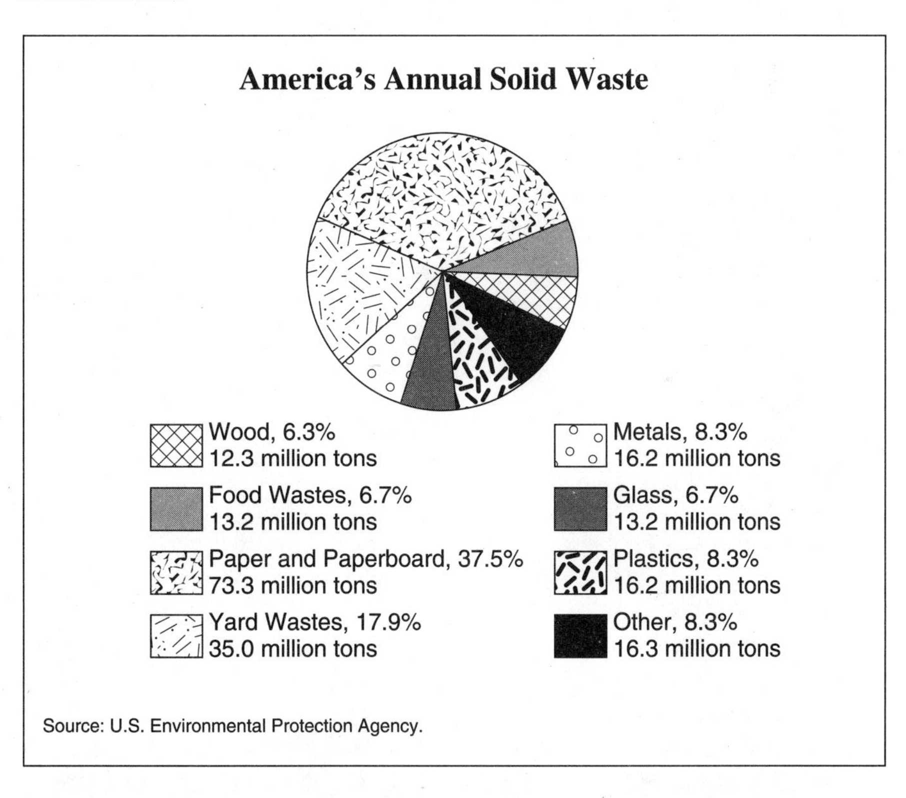

Source: U.S. Environmental Protection Agency.

EPA's 1991 regulations, issued under the Resource Conservation and Recovery Act (RCRA), set location, design, operating, and closure standards, as well as clean-up requirements for existing contamination. Importantly, the new regulations also set financial assurance criteria, requiring owners/operators of landfills to demonstrate their ability to finance required monitoring and other follow-up activities for 30 years following closure of a landfill.

Given the realities of state-of-the-art landfilling and the new federal requirements governing municipal landfills, urban areas have a compelling opportunity to bring down their total solid waste management costs with aggressive recycling and composting. In fact, some cities have already accomplished this goal.

Opponents of recycling claim that recycling services drive up costs. This is not necessarily the case. The value of recyclable commodities will improve with increased demand brought about primarily by content legislation and progressive industry target recovery goals. Consequently, when the recession ends, the market value for commodities will be improved.

> ***"We need to build a national MRF [material recovery facility] infrastructure."***

It comes down to this: All the best collection and processing strategies in the world are useless without comparable success in developing the end-use markets for recyclable commodities.

Are we on track toward creating viable markets that will support aggressive waste recovery? This, unfortunately, is where we are having the most trouble getting a foothold. For solid waste recycling to succeed as a national strategy, a decade of market development infrastructure work lies ahead, and we have three major hurdles to clear.

First, we have neither a national strategy nor enough federal governmental leadership on this issue. Second, some states, like Washington and New York, have developed successful market development programs, but most have not. And third, while some industries are actively stimulating the recovery of their products, other industries are not.

State Governments' Role

Because of slow progress at the federal level, the mantle of leadership falls on state governments. They have a unique opportunity to include market development for recyclable commodities as part of their ongoing commerce development role.

State governments do an effective job of marketing everything from tourism to produce, and Washington and New York have demonstrated that by making market development for recyclable commodities a priority, they can attract major new recycling industries. These industries offer new economic development opportunities and employment opportunities in addition to end-use markets for local government's recyclable commodities. Washington, for example, now has enough mill capacity for all waste newspaper generated in the state, and a brand new recycled content phone-directory paper mill.

The states that effectively staff and fund recyclable commodity market activities and aggressively market the availability of recyclable commodities to industry will be big winners.

Private industry can also continue to expand its leadership role. More than 95 million tons of recyclable commodities were recovered in 1991, and almost every community experienced record recovery. U.S. paper mills have already spent $42 billion in the past three years re-tooling to manufacture recycled content paper. These actions on the part of major newsprint manufacturers demonstrated that recycled-content legislation has worked effectively in the interest of creating significant new demand for waste newsprint and magazines.

Most of the major commodity groups with consumer products including plastic, metals, and glass have set target recovery goals for this decade. Those are positive steps in the right direction, and most of the goals are likely to be achieved.

Industry must also dedicate itself to manufacturing recycled content products. At least 11 states now have recycled-content laws that require the manufacturers of specific products to use recycled-content material in their products. Overall, the commodity marketing job is progressing on the right track.

We are beginning to change America's behavior from a throwaway society to a conserving society. State-mandated recycling goals have encouraged millions of Americans to voluntarily participate in solid waste recycling. More sophisticated collection systems will improve recovery and cost efficiency. Giving future MRFs a larger mission in life will help us process additional waste streams.

If, over the next decade, we can make enough progress in developing markets in recyclable commodities, the end results of this nation's integrated solid waste management approaches that include recycling will be as other industrialized nations have demonstrated: enhanced resource conservation, an improved gross national product, increased trade exports, and a reduction in solid waste and environmental degradation.

Efficient Recycling Programs Can Reduce Waste

by Christopher Juniper

About the author: *Christopher Juniper is an environmental economist in Portland, Oregon.*

Controversy about how best to support the nation's recycling desires again faces lawmakers at both the state and congressional levels. Successful establishment of markets for recycled materials continues to be the Achilles' heel of successful community recycling efforts. America's ability to recycle our waste seems to be well ahead of the markets for recycled products, and the tough questions concern the role of governments in stimulating those markets.

Recycled Shoes

For example, when Julie Lewis of Portland, Oregon, sought venture capital financing for Deja Shoes, which will be made entirely from non-virgin materials, her potential investors carefully scrutinized the reliability of the suppliers of the recycled components. Without reliable suppliers, investors would balk, and the business would not be more than Lewis's dream.

While many communities have adequate waste streams to supply Deja's needs for recycled polypropylene, paper fibers, cotton and foam rubber, only a few had companies with the technological know-how and track record to assure Deja's investors and managers that they could fill Deja's needs. The result is that Deja Shoes won't be using Portland's garbage, as Lewis had planned.

Lewis received a grant to advance her recycled shoe concept from Metro, Portland's regional manager of its solid waste stream, which had expected Deja Shoes to use Portland-area garbage. However, the most reliable suppliers of recycled shoe components are scattered throughout the country. Shoe assembly

Reprinted with permission from *Buzzworm* magazine, the environmental journal. For more information or a subscription, write 2305 Canyon Blvd., Ste. 206, Boulder, CO 80302, or call (800) 825-0061.

will occur in Taiwan, where Deja had to convince the factory to recycle waste from production of other shoes and contract with collectors to remove foam rubber from office chairs headed for landfills. All these suppliers will use local waste streams, not Portland's.

Recognizing the importance of local industries that use recycled "raw" materials as part of local solid-waste solutions, states are creating incentives for recycled material manufacturers. Kansas has joined New York, Illinois and other states in offering low-interest loans and grants to help businesses develop recycled product manufacturing.

Current prices of Deja's recycled materials are roughly comparable to virgin materials today, but as technology advances, the recycled materials supply will increase and supplier competition will evolve. Lewis hopes that prices will drop, making Dejas cheaper than shoes made from virgin materials.

Efficient Recycling

Each urban area's solid waste stream has the potential of becoming a valuable resource that can contribute to community wealth if collected and reused efficiently. But like the virgin raw materials they compete with, recycled materials are likely to be shipped around the world rather than used locally. The community that collects the valuable components of solid waste most efficiently and whose entrepreneurs turn them into cost-effective raw materials for global markets will have the most profitable—and probably the most sustainable—community recycling programs.

Recent local and state government recycling approaches have focused on separating recyclable materials from waste streams and charging for garbage collection according to the amount of non-separated materials collected. Nearly 4,000 communities have created curbside recycling programs, while others are investing in the 222 built or planned "materials recovery facilities" (MRFs) that separate recyclables from garbage in a single processing facility.

MRF systems are designed to reduce the municipal solid waste headed for landfills by up to 90 percent, through mechanical sorting, often in an enclosed building. Trying to use completely unsorted garbage, however, doesn't work well, often resulting in "contaminated" resale materials. Most cities using MRFs are bringing somewhat prescreened garbage, which lowers the amount of residue to an average of 7 percent nonrecyclable materials, according to a report by Government Advisory Associates.

> ***"States are creating incentives for recycled material manufacturers."***

Many communities are making progress toward achieving the 80 percent recycling rate of consumer garbage that some believe is possible. Bowdoinham, Maine, has achieved 54 percent recovery rate of its solid waste stream, through simple landfill user fees coupled with free collection of source-separated recycled materials, according to *Recycling and Composting Options: Lessons from 30 US Communities*, by the Institute for Local Self-Reliance.

Success Stories

Recent results of many government-initiated changes have been dramatic. Florida is now recycling 21 percent of its waste stream, compared with less than 4 percent in 1989. Lincoln Park, New Jersey (population 11,000), achieved 60 percent recovery of its municipal solid waste in 1990 through requirements that commercial establishments recycle glass, aluminum, high-grade paper, newspaper and corrugated cardboard coupled with a $119 per ton dumping fee (compared to a regional average of $64 per ton).

According to the Institute for Local Self-Reliance, Seattle leads major urban areas, with 40 percent solid waste recovery, and is well on its way to meeting its 60 percent goal by 1998. Seattle has implemented a comprehensive program for commercial and household waste, including curbside recycling and yard waste collection, apartment building recycling, commercial sector paper diversion and backyard composting. The ultimate consumer recycling potential may be 80 percent, achieved by a study group of 100 households by Barry Commoner's Queens University research group.

> ***"Many communities are making progress toward achieving the 80 percent recycling rate of consumer garbage."***

But while technical advances, tougher regulations and infrastructure investments make greater recycling possible, many recycling programs have suffered from dramatic recession-caused drops in the market value of collected materials. The value of recycled aluminum dropped 39 percent in 1991, and Americans recycled about 1 billion fewer aluminum cans that year than in 1990. (Only 57 billion of 91 billion cans were recycled in 1991—a 62 percent rate.) Recycled plastics prices dropped 49 percent, paper 61 percent and newsprint 200 percent in 1991.

A survey by *Resource Recycling* magazine found that 71 percent of state recycling coordinators cited markets for recycled materials as their greatest problem. To stimulate these markets, states are taking a variety of approaches, generally around the theme of requiring recycled content in products or packaging.

At least 18 states have implemented or negotiated recycled-content standards for newsprint. By 1995, in Oregon, glass containers must have 35 percent post-consumer recycled glass content, and rigid plastic containers must have 25 percent recycled content or meet other recycling goals. California fiberglass insulation manufacturers are now required to use 10 percent post-consumer recycled glass.

Many observers feel that nationwide laws of this sort would make more sense than 50 confusing and potentially conflicting state systems. Others staunchly advocate that legislative requirements will stifle the private sector's ability to solve the problem, while creating additional regulatory drains on tax revenues.

> ***"Many observers feel that nationwide laws . . . would make more sense than 50 confusing and potentially conflicting state systems."***

Lynn Scarlett of the Reason Foundation notes that over 160 local laws were passed in 1991 regarding recycling, "with few to none accompanied by a cost/benefit analysis for the legislation." Her solution is to use garbage unit pricing, which (along with MRFs) may reduce nonrecycled waste enough to solve landfill capacity problems, while generating an adequate stream of recycled materials to support new products, such as Deja Shoes, as recycled material manufacturers gear up capacity on their own initiative. . . .

Three Requirements

Companies new or old will not invest millions of dollars into making products from recycled materials until they (and their investors) are fully confident of three requirements: The waste stream of their community will provide a reliable and inexpensive source of high quality raw materials at a predictable price; the technology (often relatively untried) to convert recycled materials into new products will work and not become quickly obsolete; and either the new "raw" materials from our waste will be able to compete with virgin materials in global markets, or consumers will demand products—such as recycled material shoes—because of individual ethics.

Led by Senator Max Baucus (Dem.-Montana), Congress has been contemplating tougher federal requirements for use of recycled materials in packaging, modeled after Oregon's landmark 1991 recycled-content law. Baucus is convinced the private sector needs a push to make the investments that are technologically possible today to allow recycled materials to compete with virgin.

The greatest push could come from consumers (and businesses) themselves, if a strong message is sent to producers that recycled products will receive Americans' dollar votes. Consumers haven't had the opportunity to choose recycled products alongside virgin material products until recently; let's hope the dollar-vote election is a landslide.

Recycling Should Not Be Mandatory

by Jerry Taylor

About the author: *Jerry Taylor is director of natural resource studies for the Cato Institute, a libertarian public policy research organization in Washington, D.C.*

Before we conclude that recycling is the "answer," perhaps we should think hard about the question.

If the question before Congress is how best to protect the environment, conserve scarce resources, and provide for landfill space, then mandatory national recycling is not the answer.

A brief examination of the case for mandatory recycling makes that abundantly clear.

No Landfill Shortage

First, we are told that recycling will help preserve scarce landfill space. We are not, however, even remotely close to running out of space for our garbage.

Despite the "garbage trucks could ring the Milky Way galaxy" rhetoric, all of the trash America will produce over the next 1,000 years could fit into a landfill 15 square miles in size.

Clearly, concerns about "drowning in our own garbage" are misplaced.

Although we aren't running out of places to put landfills, many regions are failing to site the facilities necessary to meet future disposal needs. That is a result of political gridlock stemming from the NIMBY (Not-In-My-Back-Yard) phenomenon. National recycling mandates, however, will do little to alleviate the increasing shortfall in capacity.

Recycling 50 percent of the solid waste stream by the year 2000 (a difficult if not impossible proposition), for example, would still leave 95 million tons of garbage to be disposed of annually.

In California, recycling 50 percent of the waste stream tomorrow would ex-

Jerry Taylor, "Recycling Is Not the Answer," *Roll Call*, February 24, 1992. Reprinted with permission.

haust that state's landfill capacity by 2008 instead of 1999.

Second, we are told that recycling will help protect the environment because the alternatives—landfilling our garbage or burning it in waste-to-energy facilities—represent severe environmental threats. That, however, is specious nonsense.

Minor Risks

Even the Environmental Protection Agency considers municipal solid waste landfills to be minimal health risks. According to the agency's own findings, which were based on ultra-conservative, worst-case assumptions, 83 percent of America's solid waste landfills pose a lifetime cancer risk of less than one in one million (about the same risk inherent in drinking a glass of tap water).

A full 60 percent pose less than a one in one billion health risk. Landfill design criteria recently promulgated by the EPA will reduce those minor risks even further.

The facts about incineration are similar. Using worst-case assumptions and relying on data from old, obsolete incinerators without today's standard air pollution control devices, the EPA found that air emissions from those facilities pose a cancer risk of from one in 100,000 to one in one million.

Since modern incinerators routinely remove more than 90 percent of all toxins from the smokestack, it follows that the true health risk from modern incinerators is less than one in one million.

> ***"Clearly, concerns about 'drowning in our own garbage' are misplaced."***

The only study to examine incinerator ash in actual monofills (as opposed to stringent laboratory analysis), a study co-sponsored by the EPA and the United States Conference of Mayors, concluded that ash leachates "were close to meeting drinking water standards."

So even if ash leachates somehow found their way into groundwater, there would be little cause for concern.

The Pennsylvania Environmental Hearing Board recently put the negligible risk posed by modern waste incinerators into perspective when it noted that "the risk of a child contracting cancer from eating one peanut butter sandwich per month for 15 years is approximately 500 times greater than the risk of contracting cancer" from the emissions of an incinerator that is being considered by Drava County, Pa.

Problems of Recycling

Contrary to popular belief, recycling is not risk-free. Most recycling processes generate large amounts of hazardous waste. In the final analysis, what's more worrisome—old newspapers buried in the ground, or the toxic sludge generated in the process of de-inking them for recycling?

Finally, we are told that recycling conserves scarce resources. That statement is true only when the economics of recycling a given material are favorable.

The price mechanism, in the final analysis, is a reflection of the relative scarcity of human and natural resources. If virgin paper is cheaper for a manufacturer to use than recycled paper, it can only mean that the resources required to produce virgin paper are relatively more plentiful than the resources necessary to provide recycled paper.

> ***"Most recycling processes generate large amounts of hazardous waste."***

Mandatory recycling under those conditions would mean that the recycling process would consume more resources—energy, labor, capital, or other materials—than would be consumed if non-recycled virgin materials were used.

"Imagine a business in which the cost of the raw materials far exceeds any revenues the firm's product can generate, and the product supply far outstrips demand. You have just grasped the economics underlying curbside garbage recycling," noted Forbes Magazine in October 1991.

The environmental lobby ignores that simple but important fact and advocates the same sort of economic central planning which ultimately led to the collapse of Eastern Europe.

Finally, let's keep in mind that the natural resources supposedly "conserved" by recycling programs are not even remotely close to depletion.

Unnecessary Recycling

For example, approximately 80 percent of all material recycled today is paper—yet recycling paper doesn't save forests. Fully 87 percent of our paper stock comes from trees that are grown as a crop specifically for paper production.

Acting to "conserve trees" through paper recycling is like acting to "conserve wheat" by cutting back on bread consumption.

In fact, to the extent that paper recycling increases, tree planting will decrease as a result of reduced demand for pulp.

We are not running out of trees or forests. America has three-and-one-half times more forest land today than it had in 1920. America is growing 22 million new acres of forest annually while harvesting but 15 million acres, for a net gain of 7 million acres each year.

And why the rush to recycle glass? Are we really running out of sand?

We are told that recycling saves energy, yet energy costs are clearly part of the economic calculations that go into the pricing mechanism. Why, then, are recycled materials still more expensive for manufacturers to use than virgin materials?

Because often ignored in the energy-savings calculations peddled by the environmental lobby is the energy necessary to deliver the recyclables to collection centers, process the post-consumer material into usable commodities for manufacturers, and deliver the processed post-consumer material to manufacturing plants.

In the final analysis, the campaign for mandatory national recycling amounts to charging into a crowded theater, turning out the lights, blocking all the exits, and shouting, "Fire!"

The result is a panicky stampede through the only door left ajar—the door with Snow White's picture on it, marked "recycling." Behind that door, however, lies a path leading to wasted resources, increased environmental stress, and economic models based on those of Eastern Europe.

Recycling Is Counterproductive

by John Shanahan

About the author: *John Shanahan is a policy analyst for the Heritage Foundation, a conservative think tank in Washington, D.C.*

Some policymakers maintain that the best way to deal with municipal solid waste is to reduce the production of such waste in the first place. The main target for reduction is packaging. These policymakers mainly seek what they call "source reduction," which means producing less of the products that later become garbage. They also stress recycling, which means reusing the materials in containers and other waste to make new products. . . .

Supporters of the Baucus Bill [proposed by Montana senator Max Baucus] see recycling as preferable to landfills as a means of dealing with solid waste. To be sure, recycling can be an efficient and environmentally responsible way to deal with waste. But it is not always the best method. In some cases recycling is extremely costly, wasting more energy and creating more pollution than the landfill alternative. Whether it is the best choice depends both on regional differences and on the item being recycled.

The Costs of Recycling

Recycling costs can mount rapidly because of collection and hauling expenses. Materials either must be collected separately and hauled in separate compartments or trucks to a recycling center, or they must be separated after collection. Washington, D.C., for instance, now requires households to separate newspapers from other trash. For this, the city must send two trucks through each neighborhood on collection day rather than one. Spokane,Washington, pays over $180 per ton for recycling its garbage; this is four times the cost of burying trash in landfills.

There are also indirect costs of recycling. Example: If households must spend

John Shanahan, "A Plain Man's Guide to Garbage: The Reauthorization of the Resource Conservation and Recovery Act," *The Heritage Foundation Issue Bulletin*, March 30, 1992. Reprinted with permission.

time to separate their trash or to wash out and return bottles, this time has a real economic cost, but it does not appear in official cost figures.

Recycling Offsets Costs

In addition, recycling some materials raises more revenue to offset extra processing and collection costs than does recycling other materials. For example, ferrous and non-ferrous metals have been recycled profitably for most of this century. Much aluminum is recycled: 54 percent of all aluminum drink cans are recycled. This is because recycling aluminum cans uses 95 percent less energy per unit of "new" aluminum that is produced than refining aluminum from aluminum ore, known as bauxite. The percentage of materials that are recycled varies widely.

The reason for this variation simply is that recycling is not always profitable. It depends on the material. The total cost of recycling plastic, for example, including the cost of collection and separation, typically is 20 percent more than the cost of producing plastics from its raw materials. This is why most communities have not tried to recycle plastic waste.

Conflicting Goals

Some advocates of recycling retort that if a material, such as plastic, cannot be recycled profitably, then a recyclable substitute should be used. Any alternative, however, will have production and recycling costs. These not only may be more expensive than the original material, but may produce larger quantities of, say, atmospheric pollution. Example: metal and glass containers are heavier and bulkier than plastic containers, so more trucks would be required for transport. These additional trucks emit pollutants into the air and use fuel.

Plastic shrink wrap illustrates some of the problems with mandating recycling. Shrink wrap is relatively inexpensive because it consumes few resources, so it tends to be used whenever it can protect products. It also does not increase waste very much because it is thin and light. Recycling plastic shrink wrap, however, would consume vastly larger resources than manufacturing from virgin material. The reason: Such small amounts of shrink wrap accumulate in households at any particular time, that enormous resources such as fuel would need to be expended to collect, separate, and process it. But if a 50 percent recycling rate (or any other rate) is mandated, then use of shrink wrap is discouraged because alternative materials that are less expensive to recycle would be substituted. Generally, these materials are much bulkier and would tend to increase total waste destined for the landfill, even if the recycling

> ***"In some cases recycling is extremely costly, wasting more energy and creating more pollution than the landfill alternative."***

rate were 50 percent. This illustrates that source reduction and recycling rate goals can be in direct conflict.

The more zealous advocates of recycling tend to overlook the wider environmental implication of the policies they seek. Policymakers, however, should consider the total effect of regulations on the economy and environment, not just in narrow sectors or small locations.

A goal of those who want to see greater source reduction or recycling in the case of paper, for instance, is to save old-growth trees. Yet old-growth trees typically are not used for paper production. These trees are more valuable as lumber or plywood than as pulpwood for paper. In fact, some 87 percent of the trees cut for pulpwood are commercial plantation-grown softwoods. These trees are planted in rows and harvested every twenty years as a rotation crop. If paper recycling were mandated, the recycled pulpwood would displace virgin pulpwood in products. With less demand for virgin pulpwood, these small trees would be less profitable as a crop and the land would be put to alternative uses, leading to a decline in the number of trees. Explains Jerry Taylor, Director of Environmental Studies at the Cato Institute: "acting to 'conserve trees' through paper recycling is like acting to 'conserve wheat' by cutting back on bread consumption." The relationship between demand for virgin pulp and the number of trees planted is highlighted by the fact that, since 1970, pulpwood consumption has increased while the number of trees in America has increased by 20 percent.

The Case of Diapers

Recycling can have other unintended consequences. It can also increase pollution. Take the case of disposable diapers. Because they take up four times as much landfill space as worn-out cloth diapers, some environmentalists want regulations to discourage disposable diapers. Yet the production and cleaning of cloth diapers generate about twice as much air pollution and seven times as much water pollution as do disposable diapers, according to a study commissioned by the EPA. Thus policymakers must ask which is better for the environment: Polluting more air and water to reduce by a small amount the waste entering landfills or putting more material into landfills in order to save clean air and water? . . .

Most Americans would agree that resource conservation is an admirable goal. Indeed, the prudent use of resources is a fundamental building block of a capitalist society. But mandatory source reduction and recycling goals can waste scarce resources. Whenever recycling is profitable, as is the case for aluminum, energy and other resources tend to be conserved. But when recycling is unprofitable, it is usually a signal that resources are being wasted.

Recycling Plastics Is Impractical

by JoAnn Gutin

About the author: *JoAnn Gutin is a science writer and holds a Ph.D. in anthropology from the University of California, Berkeley.*

Somehow, the people who set up the "Reducing Plastics in the Wastestream" conference, which I went to a while back at my local Holiday Inn, seemed unclear on the concept. For one thing, the coffee was served in plastic-foam cups. For another, several conference organizers wore big buttons that said "No Bans," in reference to the ban on plastic foam that went into law in 1990 in Berkeley. But the strangest thing by far was introduced with great fanfare in the afternoon session—the Oui-Oui Skreen, an item made of old pool liners and available from a company in Cleveland.

A Metaphor for Plastic Recycling

The Oui-Oui Skreen, a spongy, blue plastic doily, saturated with an overpowering bubble-gum scent, is designed to be placed in urinals to prevent untidy splatter. But the Skreen serves a larger purpose, to which its inventor is oblivious: it is the perfect metaphor for everything that's wrong with the idea of recycling plastics. It's ugly as sin, the world doesn't need it, and it's disposable.

The Oui-Oui Skreen's enthusiastic manufacturer told an audience of waste-management professionals and politicians that the potential market for his product was enormous. There are tens of millions of urinals in the United States, he crowed, and every one of them needs a new Oui-Oui Skreen each month. In other words, what would have gone to the landfill as a plastic bottle in January would instead go as a urine-soaked plastic doily in May. Did I miss something?

The fact is that plastic will soon be to modern Americans what the walrus was to the Aleut or the buffalo was to the Sioux: nothing less than the basis of an entire material culture. In some form or other, we wear it, eat with it, write with it, cover our floors with it, insulate our houses with it—the list is practically endless.

Being less spiritually evolved than the Aleut or the Sioux, though, we loathe plastic instead of respect it; we're addicted to it, but make fun of it. Plastic is not just a substance, but a code word for a way of life that all right-thinking people despise. Plastic has gone in thirty years from a symbol for high technology and inventiveness to a symbol of rampant consumerism, from the space program to Cup O' Noodles.

A Schizophrenic Attitude

It is this schizophrenic attitude that makes the current struggle over what to do with plastic trash so interesting to watch. Advertising that once touted the convenience and disposability of plastic is now trying to soft-pedal convenience and backtrack on disposability. Anyone who has a TV has been exposed to a barrage of commercials announcing that plastic has been rehabilitated into a "recyclable" material.

Most environmentalists disagree passionately with the contention that plastics are, or even ought to be, recyclable. They cite chemical residues from plastics manufacturing, hint darkly about cancer-causing chemicals in plastic foam, outline the logistical problems with plastics collection—but mostly you can tell that they just hate the stuff. They hate it because it's based on fossil fuel; they hate it because people throw it away on the sidewalk; they hate, in short, the culture that it epitomizes. Talking about recyclable plastic to environmentalists is like talking to Nancy Reagan about legalizing drugs. In their view, anything that encourages the use of the despised substance is absolutely wrongheaded. Abstinence is the only solution.

Which leaves the poor consumer caught in philosophical cross fire. Given that most of us would rather not spend our remaining years doing environmental cost-benefit analyses of every purchase, it's no wonder that we've snuggled right up to the idea of recyclable plastics. We can all be forgiven for hoping that the environmentalists who urge us to quit cold turkey are reacting to Plastic, the metaphysical concept, not plastic, the schlocky but necessary substance.

> ***"A barrage of commercials announc[e] that plastic has been rehabilitated into a 'recyclable' material."***

And not only necessary but misunderstood, according to the industry. A brochure put out by Sonoco insists, in injured tones, that many people don't understand that plastic is made from natural materials.

Well, yes. Sort of. Plastic is natural in the same way that it's recyclable. If you're prepared to define "recyclable" so loosely that it means "not quite immortal," then plastic trash is, indeed, recyclable. But recycling plastic bears as much resemblance to traditional recycling as RV camping bears to backpacking. It might be a good idea if there were some sort of linguistic flag for the difference, the way "Kampground" has come to mean RV site. "RecyKling," maybe.

The Plastics Glut

The plastics glut began innocently enough, with the invention of celluloid, a cellulose-based material, in 1870. Besides being utilized as film, celluloid was widely used as a substitute for the ivory in billiard balls and the tortoise shell in jewelry. Ironically, in view of its environmental downside, there are probably still elderly tortoises and elephants who owe their longevity to the invention of plastic.

The first completely synthetic plastic was Bakelite, invented in 1909. Although you now pay twenty dollars a pop for Bakelite bangles in antique shops, it was throwaway stuff at the time. Not until well after World War II did plastic begin to come into its own, and remarkable as it seems, the full-blown epidemic of plastic packaging is hardly more than fifteen years old. In 1960, American consumers discarded a mere 400,000 tons of plastic; in 1988,14.4 million tons.

The current upsurge in public interest in recycling dates to the summer of 1987, when the sorry image of the wandering garbage barge towed by the tug *Break of Dawn* was seared onto the collective retina. Mass claustrophobia set in as we faced garbage gridlock. East Coast beaches were inundated with trash, much of it plastic: disposable hypodermics, contaminated with God knows what; disposable diapers; plastic lighters; throwaway pens; six-pack rings; two-liter bottles.

> ***"Recycling plastic bears as much resemblance to traditional recycling as RV camping bears to backpacking."***

Today, managers of city and private recycling programs all over the country report that people—at least the kind of people who generally worry about recycling—are desperate to recycle their plastic. Piles of plastic bottles sprout overnight like mushrooms around locked drop-off igloos in supermarket parking lots. The word has gotten around that plastic is recyclable.

But is it? Most people assume, given the traditional meaning of "recycling," that every old two-liter soda bottle becomes a new two-liter bottle. After all, when glass is recycled, bottles and jars are pulverized, combined with virgin raw materials, and made into more bottles and jars. Ditto for aluminum. That's not what happens in RecyKling. If we and the discarded soda bottle are lucky, it ends up as fiberfill for parkas, carpet backing, or tennis-ball fuzz. And although it would be comforting to think that old plastic milk jugs become new plastic milk jugs, they don't. Mud flaps for trucks, maybe, or plastic park benches, but not milk jugs. A plastic-foam cup can currently aspire to reincarnation as plastic-foam packing pellets. This may be metamorphosis, but it's not recycling. Or, in the unintentional oxymoron coined by the plastics industry, it's "linear recycling."

True, certain kinds of waste plastics can be collected, reduced to resins, and reshaped into plastic garbage cans, plastic license-plate holders, or plastic lumber. However, it's hard to escape the conclusion that drowning in a sea of plastic license-plate holders is only marginally preferable to drowning in a sea of old milk jugs. A cynic might say that the plastics industry is banking on consumers not figuring that out, or that it is at least pleasantly surprised that no one seems to have figured it out yet. A realist might conclude that the public at large, which isn't so queasy about the aesthetic and philosophical implications of plastic, just doesn't see it as a problem.

Unsuitable for Recycling

Why can't the most ubiquitous consumer plastics—milk jugs, soft-drink bottles, plastic bags—be recycled into new incarnations of themselves? A number of factors make this a recycling prospect from hell.

For one thing, the Food and Drug Administration requires that, for any container to be refilled with food or beverages, it must be sterilized, and most plastic can't take the heat.

For another, some plastics have the tendency—unlike glass and aluminum—to absorb minute quantities of what they contain. If some tidy householder had stored motor oil in a plastic milk jug, and that jug were later recycled into a new milk jug, traces of the motor oil might conceivably leach back into the milk.

But the big catch is that, for any kind of bottle recycling to work, plastic trash must be sorted. In what seems like a cosmic joke, the versatility that has enabled plastic to fill so many packaging niches is what makes recycling it so difficult. Glass is glass, more or less, and aluminum is aluminum, but plastic has a thousand faces.

There are two classes of plastics, the thermosets and the thermoplastics. Thermosets are rigid plastics; they don't flow when they're reheated, so for all practical purposes they're not recyclable (once a taillight, always a taillight). Thermoplastics are infinitely malleable, and there are somewhere between 500 and 1,000 different kinds of them.

> ***"Most people assume . . . that every old two-liter soda bottle becomes a new two-liter bottle."***

The catch is that, to recycle anything made of any one of these thermoplastic resins, the source material must be pure. One undetected high-density polyethylene milk jug in a load of two-liter polyethylene terephthalate soda bottles contaminates the whole batch. Somehow, it's difficult to believe that the average consumer will be motivated enough to learn to discriminate between polyvinyl chloride and polyethylene terephthalate.

So, mingled postconsumer plastics need to be sorted after collection. At this stage, sorting requires humans, who need to be paid. If plastic were an intrinsically valuable commodity, that wouldn't be a problem. However, it absolutely is not, which is why so much plastic collected for recycling is shipped to Asia, where labor is cheap.

The Bottom Line

Which brings us to what most recyclers agree is the bottom line of the problem with recycling plastic: the Bottom Line. Given the current, if temporary, abundance of fossil fuel, consumer plastics are cheap cheap cheap to manufacture. It is, in fact, cheaper to make plastic, from both the standpoint of raw materials and the energy consumed in manufacturing, than glass. Ironically, the reason is that plastic itself is the product of a sort of recycling process. The resins that are the raw material of modern plastics are a by-product of petroleum refining. If you can suspend judgment of the end result, that sounds like the quintessential straw-into-gold scenario.

Recycling scrap from large manufacturing operations, however, has been going on since the dawn of the plastics age. Observes Robin Boone, longtime Oakland recycling activist and board member of numerous public and private recycling concerns, "We have a saying in this business: A big enough pile of anything is worth something." Since industrial-scale manufacturing inevitably generates big piles of stuff, recycling it has always made bottom-line sense. In 1988, somewhere between 3.5 and 5 billion pounds of industrial plastic was recycled.

"The versatility that has enabled plastic to fill so many packaging niches is what makes recycling it so difficult."

But the problem is postconsumer, not postindustrial plastics. Amassing a pile of plastics big enough to be worth something costs more than making it in the first place. It's almost impossible to escape the conclusion that recycling plastic is difficult and nearly pointless.

The Green Aura

Industry executives, though caught by the current environmental furor, are unlikely to commit corporate hara-kiri. They are instead scrambling to present themselves as good citizens. But what if being a good citizen means spending ruinous amounts of stockholders' cash? In such cases, the best strategy is sometimes to spend a lot of money on advertising and public relations. Spend as much money, in other words, to advertise the research you're doing as you spend funding the research itself.

Another tack is to fund pilot RecyKling programs. At best these may actually help solve the problem, and at worst they generate a lot of positive publicity. In setting up these green-oriented projects, the company also benefits internally: employees, who are after all just folks with kids and ozone guilt like the rest of us, get a chance to feel good.

But such programs don't cut much ice with environmentalists. Kathy Evans, director of recycling for the Ecology Center and the Berkeley Buyback, summed up recyclers' position: "Styrofoam recycling is a joke. It amounts to spending money to collect large quantities of a substance that's mostly air, spending more money and fossil fuel to take it someplace, spending more money to wash it, then spending more money to shape it into something else that's mostly air. That doesn't make any economic sense."

At present, companies are paying for these Recycling programs themselves. Industry spokespersons maintain that, when small, local recycling plants, which will reduce transportation costs, are set up, recycling plastic will become economically feasible. Until that time, subsidies from either government or industry will be necessary. Why, they ask, should plastic be denied the boost given to glass, paper, and aluminum? Is it being discriminated against because the public sees it as a "nonnatural" material?

Destination: Waste Disposal

What happens to plastic that isn't RecyKled? If not littering a beach or a vacant lot, it's on the way to a landfill or an incinerator. In a sense, landfills have served us the way the portrait served Dorian Gray, by allowing developed countries to cavort in an orgy of consumerism and be spared any glimpse of the consequences—until the day of reckoning, which is either yesterday or tomorrow, depending on where you live.

Because each American generates somewhere between one thousand and three thousand pounds of trash a year, we're running out of space. More accurately, some places are running out of cheap space. Most landfills are supposed to hold between ten and thirty years' worth of solid waste. The problem is that few new ones are opening, because of local opposition.

Where does plastic fit into the garbage deluge? As the Plastics Institute of America never wearies of pointing out, the bulk of what goes into landfills, by weight, is paper, yard waste, glass, and metal. But when it comes to filling up landfill, weight isn't the issue. According to environmental sources, plastics account for over 20 percent of the wastestream by volume, and show no sign of decreasing. Plastic waste doubled in weight between 1976 and 1984 and is projected to reach 75 billion pounds by the year 2000.

"It's almost impossible to escape the conclusion that recycling plastic is difficult and nearly pointless."

And that plastic will be there for a long, long time. The average, garden-variety squeeze bottle will last well over a hundred years. Like the erosion rates of sandstone and granite, the decomposition of plastic depends on atmospheric conditions and the type and density of the material. Exposed to rain, wind, and variations in temperature, a ketchup container might disintegrate in a century and a plastic grocery bag in a year. In a nice sanitary landfill, particularly one in the arid West, both can reasonably aspire to synthetic immortality.

"Styrofoam recycling is a joke. It amounts to spending money to collect large quantities of a substance that's mostly air."

If you can't successfully bury plastic, can you burn it? That would really be a RecyKling coup: take an intrinsically worthless by-product of petroleum, then turn it into something putatively useful; when the useful object is discarded, incinerate it and create energy. It's an appealing vision—more straw into gold—and high-tech waste-to-energy plants have become widespread in Europe and Asia. Incineration was one disposal method promoted early on by plastics manufacturers, who observed with satisfaction that plastic burned twice as hot as Wyoming coal.

But incineration is fraught with problems, both those inherent in developing technologies and those inherent in burning anything. As everybody in the developed world must be heartily sick of hearing by now, burning fossil fuel results in the emission of carbon dioxide, the chemically innocuous but atmospherically insidious greenhouse gas. Plastic is fossil fuel once removed; on that basis alone, burning it is a bad idea.

Worse, many plants do what's known as "mass burns" of unsorted trash, which includes batteries, leftover paint, old refrigerators, and plastic waste. The ash that results may be full of heavy metals, and the fumes, expensive scrubbers notwithstanding, contain minute amounts of dioxins and furans.

In other words, it begins to look as though the least environmentally costly strategy, despite the appeal of plastics RecyKling as a concept, is source reduction. In other words, boring old abstinence.

In a way, packaging everything in plastic is like making hollandaise sauce: There's a dilemma about the by-product. When you make hollandaise sauce, you confront the problem of egg whites. You could throw them out, but there are so many hungry people. On the other hand, what would they do with a bowl of egg whites? You could use them up, but then you'd spend time and energy making things, like meringues, that you don't really want or need in the first place, and will probably wind up throwing away.

It's the same deal with postconsumer plastics: it is technically possible to reuse them, but is it worth it? I've given up hollandaise sauce, so I guess I can give up plastics.

Boy, the nineties are going to be the pits.

Urban Recycling Has Failed

by David M. Rothbard and Craig J. Rucker

About the authors: *David M. Rothbard and Craig J. Rucker are the president and executive director, respectively, of the Committee for a Constructive Tomorrow, a Washington, D.C., public interest group that promotes free-market solutions to consumer and environmental concerns.*

A city bus rumbles down a crowded street with the words "Recycle Now!" splashed across its sides in neon green. Around the corner, a young mother packs her groceries into a reusable canvas bag that bears the message, "For the Sake of Our Kids, Recycle." And across town at the local branch of the First National Bank, a huge sign hangs in the lobby encouraging customers to "Eat Right, Exercise Regularly, and Recycle" (not to mention "Invest in CDs").

Welcome to Anytown U.S.A. Almost out of nowhere, recycling has exploded onto the American scene. Once the domain of a few penny-wise households with extra garage space and some ambitious Cub Scouts with red wagons, this simple activity has now become a national obsession for millions concerned with "Saving the Planet."

More Popular than Voting

Recycling is so ubiquitous these days, in fact, that Jerry Powell, a well-known environmentalist who edits *Resource Recycling* magazine, quipped that recycling is more popular than democracy. After all, he explained, more Americans separate trash than vote in presidential elections.

A quick look at the facts lends credence to Powell's claim. In 1990 alone, more than 140 recycling laws were passed in 38 states. And curbside recycling programs, which in 1989 were being conducted in just 600 U.S. communities, have now ballooned to over 4,000 towns and cities across the land. As one might expect, all of this has had an enormous impact on the national recycling

David M. Rothbard and Craig J. Rucker, "Recycling: Green Panacea or Municipal Nightmare?" Position paper from the Committee for a Constructive Tomorrow. Reprinted with permission.

rate, driving it up from 13% in 1988 to a hefty 17% in 1992. This means that of the 180 million tons of garbage generated annually in America, we are recycling roughly one-third more today than we did in 1988.

This would all be just grand if recycling were truly the best solution for our garbage woes. But with reports of old newspapers piled sky-high in New Jersey warehouses, mountains of discarded bottles on the docks of Seattle, and haulers in Minneapolis incinerating recyclable materials they have no more room to store, it must be asked whether recycling is really all it's cracked up to be, and if not, why it is being touted by environmentalists as a requirement for saving Mother Earth.

"Curbside recycling programs . . . have now ballooned to over 4,000 towns and cities across the land."

Most people believe recycling is achieved when they put their newspapers out on the curb for pickup, or drop off a few cans and bottles at their local collection center. What they don't realize, however, is that recycling is only completed *after* those same papers, cans, and bottles are turned into new products and *actually sold in the marketplace*. Therefore, the mere act of throwing something into a recycling bin is no different than throwing it into a landfill if it is not used again. And it is precisely this reality that is crippling recycling programs around the country.

A Problem of Supply and Demand

The problem, simply put, is that the supply of recycled goods in America is far outpacing the demand. This is mainly because the preponderance of recycling programs in the U.S. has created a glut of recycled materials, as can be seen in their bargain basement prices. Scrap paper, which in 1988 sold for $120 a ton, is now selling for $30 a ton. Recycled glass, which enjoyed prices at the $160 a ton level in the mid 1980s, presently fetches only $90 a ton. Even aluminum, the most profitable of all recycled materials, is worth only half the value it held just a few years ago.

Since the profitability of recycling programs depends so heavily on the capacity of communities to sell the finished products at a reasonable price, this weak demand has raised the overall price of recycling and severely hurt its ability to compete with other waste-management options—namely landfilling and incineration.

Nationally, some 67% of our trash is disposed of in landfills at an average cost of $35 a ton. An additional 16% is burned in waste-to-energy incinerators at an approximate cost of $55 a ton. Recycling costs, on the other hand, vary widely, but often exceed $100 a ton. In Rhode Island, for example, communities are paying a burdensome $180 a ton for recycling while in the Windy City, Chicagoans are being strapped with an outlandish charge of between $625 and

$1,100 a ton. So costly have recycling programs become that many cities are looking to bail out.

The New York Debacle

Most prominent of these is New York. Responding to the cries from environmental advocacy organizations such as the New York Public Interest Research Group, the NYC Department of Sanitation (DOS) embarked on an intensive citywide recycling program in January 1989. The DOS set an ambitious recycling rate of 25% as its goal and sought to achieve it within five years. An impressive level of funding was also spent on the initiative as some $80 million, or roughly 12% of the department's budget, was invested to ensure the program's success.

But problems began to arise almost immediately. First was the apparent oversight of what to do with the growing supply of materials the DOS was collecting. There were no facilities constructed to process the recyclables—nor were there markets developed to sell them. And while the law required the construction of ten facilities to handle the flow of materials, after the first three years of operation none had been developed.

All this had the effect of driving the costs skyward. The $80 million the DOS invested in the program was quickly swallowed up and by the end of the first year the actual expenses totaled $127 million. The projected costs to process recycled materials also rose from $65 a ton to over $300 a ton. This trend continues to escalate with city officials now estimating total program costs to soon exceed $250 million a year.

"So costly have recycling programs become that many cities are looking to bail out."

Faced with these problems, Mayor David Dinkins decided to throw his full support behind the construction of a massive waste-to-energy incinerator in Brooklyn's Navy Yard. Despite some angry protests from environmentalists who still support the recycling program, the incinerator is presently scheduled to be built in 1996, and the city will begin hauling a large portion of its trash there before the end of the decade.

Seattle: A Model Program?

New York's experience, while dramatic in its failure, is by no means unique. Many solid waste programs around the country are fast going broke and changing direction. Perhaps even more troubling for recycling proponents, though, are the problems beginning to crop up in supposedly "successful" recycling initiatives. Seattle represents perhaps the best case in point.

Originally started by the Seattle Solid Waste Utility (SWU) as a response to the closing of the municipality's only two landfills, the city's program initially gave not only environmentalists, but even skeptics, a reason to smile. When the SWU set a recycling goal of 60% by 1998, most observers only scoffed. But after just four years into the program (which originally started in 1988), Seattle had already achieved an impressive recycling rate of well over 30%. Indeed, the program was so successful that many regarded it as the finest example of curbside recycling in the nation.

But just as with other cities experimenting with recycling, Seattle soon discovered Murphy's Law. The rapid collection of recyclables, while indeed impressive, had not been followed by any increased demand for their use. In fact, it had created a glut that drove prices downward and cost the city more than the SWU had anticipated. Glass prices dropped from $40 a ton to under $10, and the city, which used to receive $10 a ton for selling paper, is now paying $5 a ton to get rid of it. To make matters worse, the primary regional landfill—a county dump in use since the city closed its own facilities—has been replaced with another one that is many miles east of the Cascade Mountains. This has had a notable impact on the overall costs of the program as trash haulers are now paying significantly higher costs in transportation fees. Thus, even in Seattle, recycling is proving to be both troublesome and expensive.

Bureaucratic and Logistical Troubles

Although costs are undoubtedly the biggest factor forcing many cities to reassess the effectiveness of their recycling programs, bureaucratic mismanagement is making other cities take a second look as well. Chief among them is Los Angeles, where the city's Bureau of Sanitation has been plagued with a myriad of operational difficulties in trying to implement its program.

First proposed by Mayor Tom Bradley in 1989, the LA recycling program was designed to comply with the State of California's mandate to reduce the city's solid waste stream by 25% in 1995, and 50% by the year 2000. Not content with merely implementing the law in the timeframe provided, the mayor proposed to set a stricter standard for Los Angeles—the city was to achieve a 30% recycling rate by 1992. Toward that end, the city government budgeted an impressive $170 million a year, an amount more than adequate to fund such an effort.

> ***"Many solid waste programs around the country are fast going broke and changing direction."***

Despite the city's good intentions, however, LA never met its goal of 30%. Indeed, neither did it meet a goal of 25% . . . 20% . . . 15% . . . or even 10%. According to the Bureau of Sanitation, by 1992 only 3% of the total trash stream the department collected was being diverted to the city's recycling program. The reason: internal logistical problems. Delays resulting from broken down vehicles, a municipal hiring freeze, inadequate numbers of trash cans, and a general lack of direction from the City Council have set the LA program back some three years behind schedule.

Other cities are reporting similar logistical horror stories. In Cincinnati, the city failed to provide enough storage space for the influx of recycled paper coming from its collection program. Similarly in the District of Columbia, tons of recycled newspapers are rotting in a pit with apparently nowhere to go. And in Salt Lake City, a great deal of time and manpower is being wasted as haulers truck garbage literally hundreds of miles to collection centers to process recyclables that could be more cheaply thrown into a local landfill.

Environmentalists Defend Recycling

Despite its many logistical and financial drawbacks, green activists have been able to sell the public on recycling by warning about the need to save "precious" natural resources and raising alarm about the safety of landfills and incinerators. Recycling, they claim, protects vast stretches of our forests from being pillaged and does not pump toxic chemicals into our air and water. So who can argue with recycling, even if it costs a few more bucks, they contend.

> ***"Los Angeles . . . has been plagued with a myriad of operational difficulties in trying to implement its program."***

Actually, this argument falls flat on several fronts. To begin with, paper consumption is not wiping out our forests. According to the American Forest Council, we have more trees today than we did 70 years ago, with at least two trees being planted for every one cut. In addition, 87% of our paper comes from trees specifically planted for that purpose. As Jerry Taylor of the respected Cato Institute noted, "Acting to conserve trees through paper recycling is like acting to conserve wheat by cutting back on bread consumption." And as for recycling bottles, America will run out of glass when we run out of sand.

Secondly, the alleged public health advantage recycling holds over its competitors is a myth. The EPA asserts that virtually all of America's landfills pose nearly zero risk and that the likelihood of contracting cancer from one is 30 times less than the chance of being struck by lightning. Concerning waste-to-energy plants, the Pennsylvania Environmental Hearing Board reported that a child is 500 times more likely to get cancer from eating one peanut butter sandwich a month than from the emissions of a local incinerator. Frankly, it is hard to find credible toxicologists or epidemiologists who are concerned about the safety of modern landfills and incinerators.

Thirdly, recycling itself is by no means void of all safety and health concerns. Many recycling centers are littered with broken glass and bottles requiring workboots, gloves, and hardhats to ensure worker safety. In addition, sharp objects occasionally find their way into the recycled materials, not only putting the workers at risk, but also the general public. Such was the case with the city of Syracuse, NY, where in 1991 city officials refused to accept shipments from a nearby processing center in the town of Bethlehem because dangerous hypodermic needles were found in the recycled materials. Finally, some recycled products such as paper plates and cups are suspected of being a health hazard. The Food and Drug Administration not long ago expressed "concern" about the presence of "chemical or microbiological contaminants" found in these recycled products—making them particularly dangerous considering the high probability of directly contaminating the food people consume.

Contempt for Consumerism

Given the exorbitant cost of recycling and the fact it achieves only a minimal public service, one must wonder why environmentalists continue to embrace it so vehemently. The answer may well lie in their apparent contempt for what they view as America's wasteful, consumer-oriented society, as illustrated by the comments of Worldwatch Institute researcher Alan Durning. "The American middle class has been on a 50-year shopping spree that threatens to bankrupt the ecology of the planet," he remarked, "and our over consumption rivals overpopulation as a global threat."

This philosophy may further explain why green leaders have been so anxious to salvage recycling programs by pushing expensive laws—such as the ones recently passed in Oregon, California, and Florida—that create markets for recycled materials by mandating their use.

In sum, America's recycling craze may well serve the political aims of the environmental movement and put money into the pockets of some trash companies, but whether or not it does anything to solve our nation's garbage problem, much less "Save the Earth," is dubious at best.

Glossary

acid rain Acid pollution that falls to the ground mixed with precipitation.

carbon dioxide (CO_2) A natural gas present in the air. It is absorbed by plants in photosynthesis and contributes to the **greenhouse effect** when produced by burning **fossil fuels** and trees.

carbon monoxide (CO) An odorless, colorless, poisonous gas produced when carbon burns with insufficient oxygen.

carcinogen A substance that tends to produce a cancer.

chlorine A greenish yellow, poisonous gas that is highly irritating to the respiratory system. It is commonly used for water purification and in making bleach. It is suspected that a single chlorine atom in the atmosphere can dissolve up to one hundred thousand **ozone** molecules in the upper atmosphere by breaking off one of ozone's three oxygen atoms. The resultant oxygen molecule, unlike ozone, cannot reflect **UV radiation**.

Chlorofluorocarbons (CFCs) Synthetic compounds composed of atoms of **chlorine**, **fluorine**, and carbon. CFCs are used commercially as coolants and in other products and are potentially damaging to the earth's **ozone layer**.

DDT A pesticide banned in the United States but used elsewhere around the world.

ecosystem A community of animals and plants and the environment that sustains it.

effluent Any liquid flowing out of a process or container. The term usually describes the discharge of a liquid pollutant.

EPA Environmental Protection Agency. The EPA is the federal agency responsible for controlling pollution in the areas of air, water, solid waste, pesticides, radiation, and toxic substances.

fluorine A yellowish, toxic, highly corrosive gas. It is the most chemically reactive nonmetallic element and reacts with most organic and inorganic compounds.

fossil fuels Fuels such as coal, oil, and natural gas, which are formed in the earth from plant or animal fossils.

Freon The trademark name of a fluorinated **hydrocarbon** used as a refrigerant.

green A term used to describe individuals, groups, or companies that work to protect the environment.

greenhouse effect A warming effect, such as that produced by a greenhouse, caused by **greenhouse gases**.

greenhouse gases Atmospheric gases such as **carbon dioxide**, water vapor, **methane**, and **chlorofluorocarbons** that hold in and reflect back infrared energy from the earth's surface, thus heating the atmosphere.

hydrocarbons Any compounds containing only hydrogen and carbon, such as **methane**. Useful as fuels and solvents, hydrocarbons contribute to **ozone pollution**.

leachate Liquids that have filtered down through the waste in a landfill, becoming contaminated with toxins and pollutants.

methane An odorless, colorless, and extremely flammable gas produced by the decomposition of landfill waste.

Montreal Protocol Montreal Protocol on Substances That Deplete the Ozone Layer. A 1987 international agreement to reduce **CFC** production by 50 percent from 1986 levels by 1999.

nitrogen oxides (NO_x) Harmful gases emitted by automobile engines, coal-fired power stations, fertilizers, and other sources.

ozone (O_3) A molecule of oxygen having three atoms rather than two.

ozone layer A natural protective layer in the atmosphere that prevents much of the sun's harmful ultraviolet light from reaching the earth's surface.

ozone pollution Air pollution, also known as **smog**, that is formed by the reaction of sunlight with **hydrocarbons**, **nitrogen oxides**, and **VOCs**.

PCBs Polychlorinated biphenyls. Poisonous industrial chemicals used to manufacture paints, pesticides, plastics, and other products. PCBs are a suspected **carcinogen** and contaminate bays, lakes, and rivers via sewage systems and through industrial dumping.

RCRA Resource Conservation and Recovery Act. Legislation passed in 1976 that established a framework for managing the disposal of hazardous wastes and municipal solid waste.

recycling The process of making a product from used materials such as aluminum, glass, and paper. The recycling "loop" is completed only when the new product is sold.

scrubbers Pollution control devices in incineration facilities that neutralize acid gases by mixing them with a lime solution.

smog A brownish haze of air pollution that often forms over urban areas.

sulfur dioxide (SO_2) A malodorous gas produced by both volcanoes and industrial activity, especially fossil-fuel burning. SO_2 is considered a primary cause of **acid rain**.

Superfund The common name for the 1980 Comprehensive Environmental Response, Compensation, and Liability Act (CERCLA). Superfund relies on a tax on petroleum and chemical raw materials as a means of financing cleanups of America's hazardous waste disposal sites.

toxicity The quality of being poisonous.

UV radiation Ultraviolet radiation from the sun. Prolonged direct exposure to it can damage the human immune system, cause cataracts, and increase the incidence of skin cancer.

VOCs Volatile organic compounds that generate harmful emissions and contribute to **ozone pollution**.

waste streams All waste coming into, through, or out of a facility.

Bibliography

Books

American Council on Science and Health	*Issues in the Environment.* New York: American Council on Science and Health, 1992.
Ronald Bailey	*Eco-Scam: The False Prophets of Ecological Apocalypse.* New York: St. Martin's Press, 1993.
Robert C. Balling Jr.	*The Heated Debate: Greenhouse Predictions Versus Climate Reality.* San Francisco: Pacific Research Institute for Public Policy, 1992.
Lester R. Brown et al.	*State of the World 1994.* New York: W.W. Norton, 1994.
Gary C. Bryner	*Blue Skies, Green Politics: The Clean Air Act of 1990.* Washington, DC: Congressional Quarterly Inc., 1993.
Robert D. Bullard	*Dumping in Dixie: Race, Class, and Environmental Quality.* Boulder, CO: Westview Press, 1990.
Frances Cairncross	*Costing the Earth: The Challenge for Governments, the Opportunities for Business.* Boston: Harvard Business School Press, 1992.
Helen Caldicott	*If You Love This Planet: A Plan to Heal the Earth.* New York: W.W. Norton, 1992.
Jennifer Carless	*Taking Out the Trash: A No-Nonsense Guide to Recycling.* Washington, DC: Island Press, 1992.
Center for Investigative Reporting and Bill Moyers	*Global Dumping Ground: The International Traffic in Hazardous Waste.* Washington, DC: Seven Locks Press, 1990.
Thomas W. Church and Robert T. Nakamura	*Cleaning Up the Mess: Implementation Strategies in Superfund.* Washington, DC: The Brookings Institution, 1993.
David E. Cooper and Joy A. Palmer, eds.	*The Environment in Question: Ethics and Global Issues.* New York: Routledge, 1992.
Mark H. Dorfman, Warren R. Muir, and Catherine G. Miller	*Environmental Dividends: Cutting More Chemical Wastes.* New York: INFORM, 1992.
Paul R. Ehrlich and Anne H. Ehrlich	*Healing the Planet: Strategies for Resolving the Environmental Crisis.* Reading, MA: Addison-Wesley, 1991.
Jon Erickson	*World Out of Balance: Our Polluted Planet.* Blue Ridge Summit, PA: Tab Books, 1992.

Murray Feshbach and Alfred Friendly Jr.	*Ecocide in the USSR: Health and Nature Under Siege*. New York: Basic Books, 1992.
Jack Fishman and Robert Kalish	*Global Alert: The Ozone Pollution Crisis*. New York: Plenum Press, 1990.
Al Gore	*Earth in the Balance: Ecology and the Human Spirit*. Boston: Houghton Mifflin, 1992.
Ronald E. Gots	*Toxic Risks: Science, Regulation, and Perception*. Boca Raton, FL: Lewis Publishers, 1993.
K. A. Gourlay	*A World of Waste: Dilemmas of Industrial Development*. London: Zed Books, 1992.
Nancy Sokol Green	*Poisoning Our Children: Surviving in a Toxic World*. Chicago: Noble Press, 1992.
Christoph Hilz	*The International Toxic Waste Trade*. New York: Van Nostrand Reinhold, 1992.
Joel S. Hirschhorn and Kirsten U. Oldenburg	*Prosperity Without Pollution: The Prevention Strategy for Industry and Consumers*. New York: Van Nostrand Reinhold, 1991.
Walter J. Karplus	*The Heavens Are Falling: The Scientific Prediction of Catastrophes in Our Time*. New York: Plenum Press, 1992.
Mark Lappé	*Chemical Deception: The Toxic Threat to Health and the Environment*. San Francisco: Sierra Club, 1992.
Rogelio A. Maduro	*Holes in the Ozone Scare*. Washington, DC: Twenty-First Century Science Associates, 1992.
Daniel Mazmanian and David Morrell	*Beyond Superfailure: America's Toxics Policy for the 1990s*. Boulder, CO: Westview Press, 1992.
National Academy of Sciences	*One Earth, One Future: Our Changing Global Environment*. Washington, DC: National Academy Press, 1992.
Political Economy Research Center	*PERC Resource Book on Pollution, Trade, and Aid*. Seattle: Knowledge Network, 1992.
William Rathje and Cullen Murphy	*Rubbish! The Archaeology of Garbage*. New York: HarperCollins, 1992.
Dixy Lee Ray and Lou Guzzo	*Environmental Overkill: Whatever Happened to Common Sense?* Washington, DC: Regnery Gateway, 1993.
Donald Rebovich	*Dangerous Ground: The World of Hazardous Waste Crime*. New Brunswick, NJ: Transaction Books, 1992.
Rolling Stone	*The Rolling Stone Environmental Reader*. Washington, DC: Island Press, 1992.
Glenn E. Schweitzer	*Borrowed Earth, Borrowed Time: Healing America's Chemical Wounds*. New York: Plenum Press, 1991.
Seth Shulman	*The Threat at Home: Confronting the Toxic Legacy of the U.S. Military*. Boston: Beacon Press, 1992.

Bruce Smart, ed.	*Beyond Compliance: A New Industry View of the Environment.* Washington, DC: World Resources Institute, 1992.
United Nations Environment Programme	*Saving Our Planet: Challenges and Hopes.* New York: United Nations Environment Programme, 1992.
Anthony B. Wolbarst, ed.	*Environment in Peril.* Washington, DC: Smithsonian Institution Press, 1991.
Nancy Wolf and Ellen Feldman	*Plastics: America's Packaging Dilemma.* Washington, DC: Island Press, 1991.

Periodicals

Peter Asmus	"Saving Energy Becomes Company Policy," *The Amicus Journal*, Winter 1993.
The Bulletin of the Atomic Scientists	Special issue on the global warming debate, June 1992.
William K. Burke	"The Booming Business in Waste," *E: The Environmental Magazine*, May/June 1993.
William K. Burke	"The Toxic Truth," *In These Times*, May 3, 1993.
Ruth N. Caplan	"Growing the Economy Green," *Christian Social Action*, March 1993. Available from 100 Maryland Ave. NE, Washington, DC 20002.
ChemEcology	"Toxic Emissions Down One-Third Since 1987," June 1992. Available from Chemical Manufacturers Association, 2501 M St. NW, Washington, DC 20037.
Consumers Digest	"Recycled or Recyclable?" January/February 1993.
Gregg Easterbrook	"A House of Cards," *Newsweek*, June 1, 1992.
EPA Journal	Special issue on minorities and the environment, March/April 1992. Available from PO Box 371954, Pittsburgh, PA 15250-7954.
Murray Feshbach	"Ecology's Eastern Front," *U.S. News & World Report*, July 20, 1992.
Christine Gorman	"Getting Practical About Pesticides," *Time*, February 15, 1993.
Rob Gurwitt	"Markets for Recycling: The Role of the States," *Governing,* January 1993.
Rose Gutfield	"In Many Areas, There's Improvement in the Air," *The Wall Street Journal*, October 16, 1992.
James P. Hogan and Frederik Pohl	"Ozone Politics," *Omni*, June 1993.
William F. Jasper	"Environmental Police State," *The New American*, May 17, 1993. Available from 770 Westhill Blvd., Appleton, WI 54915.

John W. Johnstone — "Chemical Industry Cleans Up Its Act," *Forum for Applied Research and Public Policy*, Spring 1993. Available from 1005 Mississippi Ave., Davenport, IA 52803.

Brian Keating and Dick Russell — "Inside the EPA," *E: The Environmental Magazine*, July/August 1992.

Florentin Krause — "The Greenhouse Dividend," *Earth Island Journal*, Fall 1992.

Fred Krupp — "Business and the Third Wave," *Vital Speeches of the Day*, August 15, 1992.

Dwight R. Lee and Robert L. Sexton — "Pollution Can Be Controlled with Less Government Regulation," *USA Today*, March 1993.

Robert W. Lee — "Superfund or Superfraud?" *The New American*, May 17, 1993.

Lisa Y. Lefferts — "Too Many Risks in the Sea," *E: The Environmental Magazine*, January/February 1993.

Jon R. Luoma — "Healing the Earth?" *Discover*, January 1993.

Jonathan Marshall — "Paying for Pollution," *Reason*, April 1993.

David Moberg — "Clinton's Burning Issue," *In These Times*, May 3, 1993.

Peter Montague — "Polluters and Their Friends," *Lies of Our Times*, June 1993.

Will Nixon — "What Does EPA Stand For?" *Utne Reader*, November/December 1992.

Frank Popoff — "Companies Change Course," *EPA Journal*, September/October 1992.

Boyce Rensberger — "The O_3 Solution," *The Washington Post National Weekly Edition*, April 26-May 2, 1993.

James Ridgeway — "Toxic Waste Syndrome: Is It Business as Usual at the EPA?" *The Village Voice*, March 23, 1993. Available from 36 Cooper Square, New York, NY 10003.

Lynn Scarlett — "Don't Buy These Environmental Myths," *Reader's Digest*, May 1992.

S. Fred Singer — "Benefits of Global Warming," *Society*, March/April 1992.

Michael Specter — "Sea-Dumping Ban: Good Politics, but Not Necessarily Good Policy," *The New York Times*, March 22, 1993.

Valerie Taliman — "Nuking Native America," *Third Force*, March/April 1993.

Joanna D. Underwood and Bette Fishbein — "Making Wasteful Packaging Extinct," *The New York Times*, April 4, 1993.

Robert M. White — "Regulations Shouldn't Be Relics," *Technology Review*, May/June 1993.

John E. Young — "It's Time to Toss Out the Throwaway Habit," *USA Today*, September 1992.

Organizations to Contact

The editors have compiled the following list of organizations that are concerned with the issues debated in this book. All have publications or information available for interested readers. For best results, allow as much time as possible for the organizations to respond. The descriptions below are derived from materials provided by the organizations. The list was compiled upon the date of publication. Names, addresses, and phone numbers of organizations are subject to change.

American Council on Science and Health (ACSH)
1995 Broadway, 2d Fl.
New York, NY 10023-5860
(212) 362-7044

The council is an association of scientists and doctors concerned with public health. It believes the environmental crisis is exaggerated and works to calm the fears of Americans who believe that their air, water, and food are contaminated. ACSH publishes the magazine *Priorities: For Long Life and Good Health* and *Media Updates* quarterly.

Chemical Manufacturers Association (CMA)
2501 M St. NW
Washington, DC 20037
(202) 887-1100

CMA is a national association of chemical companies. Its activities include doing research on air and water pollution control and providing health and safety information about chemicals to the public. CMA asserts that chemical companies are making progress to reduce toxic emissions and waste. It publishes the magazine *ChemEcology* ten times a year and makes available booklets on the safe use of chemicals.

Competitive Enterprise Institute (CEI)
1001 Connecticut Ave. NW, Suite 1250
Washington, DC 20036
(202) 331-1010

CEI encourages the use of private incentive and property rights to protect the environment. It advocates removing government barriers to establish a system in which the private sector would be responsible for the environment. Its publications include the monthly newsletter *CEI UpDate* and numerous reprints and briefs.

Earth Island Institute
300 Broadway, Suite 28
San Francisco, CA 94133
(415) 788-3666

The institute works to prevent the destruction of the environment by sponsoring environmental and wildlife protection projects. It publishes the quarterly magazine *Earth Island Journal* and brochures that describe the dangers of greenhouse gases and ozone depletion.

Environmental Protection Agency (EPA)
401 M St. SW
Washington, DC 20460-0001
(202) 382-2090

EPA is the federal agency in charge of protecting the environment and controlling pollution. The agency works toward these goals by assisting businesses and local environmental agencies, enacting and enforcing regulations, identifying and fining polluters, and cleaning up polluted sites. It publishes the quarterly *EPA Journal* and periodic reports.

Environmental Research Foundation
PO Box 5036
Annapolis, MD 21403
(410) 263-1584

The foundation works with various groups and individuals, such as grassroots activists and public-interest scientists, to motivate communities to act against the dangers of all types of pollution. It specializes in information about hazardous waste and waste disposal and seeks to educate the public on their adverse health effects. The foundation's publications include the weekly newsletter *Rachel's Hazardous Waste News* and various fact sheets and reports, including *An Odor Like a Skunk Dipped in Creosote and Burned: EPA's Regulation of Commercial Hazardous Waste Incinerators.*

Greenpeace U.S.A.
1436 U St. NW
Washington, DC 20009
(202) 462-1177

Affiliated with Greenpeace International, this organization consists of conservationists who believe that verbal protests against threats to the environment are inadequate; it takes action. Among its various activities, Greenpeace monitors conditions of environmental concern such as the greenhouse effect and toxic waste dumping, and it conducts nonviolent protests against polluters. It publishes the quarterly newsletter *Greenpeace* and many books and reports, including *Global Warming: The Greenpeace Report.*

The Heritage Foundation
214 Massachusetts Ave. NE
Washington, DC 20002
(202) 546-4400

The Heritage Foundation is a conservative think tank that supports free enterprise and limited government. Its researchers criticize EPA for overregulation and believe that recycling is an ineffective method of dealing with waste. Its publications, such as the quarterly *Policy Review* and the *Heritage Lectures*, include studies on the uncertainty of global warming and the greenhouse effect.

INFORM
381 Park Ave. S.
New York, NY 10016
(212) 689-4040

INFORM is a research organization that studies practical actions for the protection and conservation of natural resources and public health. Its studies, which cover many environmental issues, have found significant progress among industries to reduce toxic waste and improve the environment. Its publications include the quarterly newsletter *INFORM Reports* and the reports *Cleaner Products—Toxics Use Reduction* and *Trading Toxics Across State Lines.*

National Recycling Coalition (NRC)
1101 30th St. NW
Washington, DC 20007
(202) 625-6406

NRC is composed of individuals and environmental and business organizations advocating the recovery, reuse, and conservation of materials and energy. It works for education, legislation, and industrial policies that will foster recycling. The coalition publishes the quarterly newsletter *NRC Connections.*

National Toxics Campaign Fund (NTCF)
1168 Commonwealth Ave.
Boston, MA 02134-4634
(617) 232-0327

NTCF is a coalition of concerned citizens, consumer organizations, scientists, and others working to solve the nation's toxic chemical and waste problems. Objectives of the fund include pressuring corporations and the military to clean up and reduce pollution. Its publications include the quarterly newsletters *Toxic Times* and *Touching Bases* and the bimonthly newsletter *Communities of Resistance.*

Natural Resources Defense Council (NRDC)
40 W. 20th St.
New York, NY 10011
(212) 727-2700

The council is a group of activist scientists, lawyers, and other citizens who work to promote environmentally safe energy sources and the protection of the environment. Its concerns include air and water pollution and toxic substances. NRDC publishes the quarterly *Amicus Journal* and the bimonthly newsletter *NRDC Newsline.*

Political Economy Research Center (PERC)
502 S. 19th Ave., Suite 211
Bozeman, MT 59715
(406) 587-9591

PERC is a research and educational foundation that advocates the use of free-market principles to protect the environment. The center believes that global warming and the environmental crisis are exaggerated. Its publications include the *PERC Resource Book on Pollution, Trade, and Aid* and the triannual *PERC Reports* newsletter.

Reason Foundation
3415 S. Sepulveda Blvd., Suite 400
Los Angeles, CA 90034
(310) 391-2245

The foundation promotes individual freedoms and free-market principles. Its researchers criticize the common belief that recycling is beneficial and declare that the dangers of ozone depletion and global warming are myths. It publishes the monthly magazine *Reason.*

Sierra Club
100 Bush St.
San Francisco, CA 94104
(415) 291-1600

The Sierra Club is a grass-roots organization with chapters in every state. Maintaining separate committees on air quality, global environment, and solid waste, among others, it promotes the protection and conservation of natural resources. It publishes the bimonthly magazine *Sierra* and the biweekly *Sierra Club National News Report*, in addition to books and fact sheets.

Society for Environmental Truth (SET)
625 N. Van Buren
Tucson, AZ 85711
(602) 790-4769

SET is an organization of academics and professionals who seek solutions to environmental problems through free enterprise and less government regulation. The society acknowledges serious pollution problems but believes that the environmental crisis is exaggerated. Its topics of research include energy, mineral resources, and air and water pollution. SET publishes *The Torch*, a bimonthly newsletter.

Worldwatch Institute
1776 Massachusetts Ave. NW
Washington, DC 20036-1904
(202) 452-1999

Worldwatch is a research organization that analyzes and focuses attention on global problems, including environmental concerns such as nuclear waste and the relationship between trade and the environment. It compiles the annual *State of the World* book and publishes the bimonthly magazine *World Watch* and the Worldwatch Paper Series, including *Clearing the Air: A Global Agenda* and *Nuclear Waste: The Problem That Won't Go Away.*

Index

Index